NON-LETHAL WEAPONS AS LEGITIMIZING FORCES?

 # 156.15

NON-LETHAL WEAPONS AS LEGITIMIZING FORCES?

Technology, Politics and the Management of Conflict

250201

BRIAN RAPPERT

University of Nottingham

Foreword by
Frank Barnaby

FRANK CASS
LONDON • PORTLAND, OR

First published in 2003 in Great Britain by
FRANK CASS PUBLISHERS
Crown House, 47 Chase Side
London N14 5BP

and in the United States of America by
FRANK CASS PUBLISHERS
c/o ISBS, 920 NE 58th Avenue, Suite 300
Portland, Oregon 97213-3786

Website: www.frankcass.com

Copyright © 2003 Brian Rappert

British Library Cataloguing in Publication Data

Rappert, Brian
  Non-lethal weapons as legitimizing forces?: technology,
  politics and the management of conflict
  1. Nonlethal weapons – Political aspects 2. Nonlethal weapons
  – Case studies 3. Conflict management – Evaluation
  I. Title
  355.8′2

ISBN 0-7146-5440-X (cloth)
ISBN 0-7146-8360-4 (paper)

Library of Congress Cataloging-in-Publication Data

Rappert, Brian.
  Non-lethal weapons as legitimizing forces?: technology, politics,
  and the management of conflict / Brian Rappert.
    p. cm.
  Includes bibliographical references and index.
  ISBN 0-7146-5440-X (cloth)
  1. Nonlethal weapons – Government policy. I. Title.

U795.R37 2003
355.8 – dc21   2003040935

Typeset in ClassGaramond, 10.5/12pt by Frank Cass Publishers
Printed in Great Britain by MPG Books, Bodmin, Cornwall

Contents

Illustrations

Foreword

On first sight, non-lethal weapons are attractive. Surely it is better to only temporarily injure someone than kill him or her. However, if misused, so-called non-lethal weapons kill. For example, as shown in Vietnam and Northern Ireland, riot-control 'incapacitating' gases such as tear gas and CN can kill if used in confined spaces. And the use of rubber bullets and stun guns has often caused fatalities.

History shows that weapons – lethal and non-lethal – often do not work in the ways their manufacturers claim. We have, for example, seen how frequently 'precision-guided' weapons, designed to reduce 'collateral damage', fail to hit their targets.

Non-lethal weapons raise important questions about how death and injury should be weighed, bringing in judgements about the value of life and the legitimacy of weapons. Most people would disapprove of the use of, for example, laser weapons to blind troops, regarding them as inhumane weapons.

One wonders about the non-lethal weapons now being researched – such as acoustic infrasound weapons, designed to nauseate and disorientate; electromagnetic pulse beams, designed to increase body temperatures to intolerable levels by heating water under the skin; and a variety of chemical agents and psychotropic drugs, to induce hallucinations, anxiety and fatigue by targeting specific cells in the brain.

In particular, the development of genetically engineered microbes and psychopharmological drugs for use as future human incapacitants will raise serious and difficult questions about the definition of biological weapons and the scope of the 1972 Biological and Toxin Weapons Convention. Just how benign will these 'non-lethal' weapons be? Not very, argues Brian Rappert in *Non-lethal Weapons and Legitimizing Forces*?

Non-lethal weapons raise important questions about the legitimacy of the use of force. These questions should be addressed in relation to future developments in national, regional and global security. Security at all levels will have to be fundamentally redefined to take into account future threats that will have evolved from a variety of global problems.

The most serious security problems will arise from the rapidly growing world population and the widening gap between rich and poor countries. This gap is likely to have increased North–South tension to such an extent that it will be greater than East–West tension ever was.

According to UNDP, a quarter of the today's world's population remains in severe poverty even though the global GDP is about $25 trillion. About a third of the people in the developing countries – 1.3 billion people – live on incomes of less than $1 a day. Nearly a billion people are illiterate. Well over a billion lack access to safe water. And nearly a third of the people in the least developed countries are not expected to survive to age 40. As populations grow it is unlikely that the gap between rich and poor countries will narrow significantly.

Security will have to be fundamentally redefined to take into account the effects of: environmental degradation; the debt crisis; the biodiversity crisis (a large fraction of existing species, possibly half, may be lost within the foreseeable future); global climate changes, brought about by global warming; and deforestation. All of these are exacerbating North–South tension and are already beginning to be seen as important elements in the security of the industrialised countries. And mounting pressures on non-renewable resources – fertile land, fish, and particularly water – are increasingly recognised as major future threats to the security of, and causes of conflict between, Third World countries, and are therefore threats to global security. These threats will become increasingly serious as time goes on.

Many of these threats to security arise from and are exacerbated by increases in population. According to demographers, the world population will probably increase from today's 5,300 million to about 7,200 million by the year 2010 and then increase to about 11,000 million by 2030. The bulk of the increase will occur in the Third World, and most of it will go into cities so that huge metropolises will be created, with populations of many tens, and even hundreds of millions of people.

Most of the inhabitants of these megacities will live in abject poverty in shantytowns with inadequate, or often no, access to medical services, safe water, food, clothing, and housing. They will have woefully inadequate supplies of energy and little access to useful education services. These conditions are just those in which urban violence and terrorism flourish.

Future terrorists and urban guerrillas will be able to acquire, through the global arms trade, sophisticated weapons at relatively low cost, including long-range ballistic missiles equipped with nuclear, chemical or biological warheads. They will also be able to deliver weapons of mass destruction by simpler methods - perhaps carried in ships.

Problems arising from a huge world population will dominate the security policies of all nations. The Earth's finite carrying capacity will almost certainly be unable to cope adequately with such large numbers of people, even though technological advances are likely to allow a somewhat higher world population than today's to be supported. Unless drastic steps are taken soon, disasters – war, pestilence, and famine – are inevitable. The potential for future violence and conflict is obvious.

The world's scientific and technical skills should be mobilised to deal with global problems. The global problems arising from the population explosion and the problems arising from poverty could be solved, or at least greatly reduced, by appropriate technology. Unfortunately, the skills needed to apply our scientific and technical skills to solving these problems are tied up by the military. About 60 million people are absorbed in military and military-related activities. Many of them are highly qualified. The most qualified are those working in military research and development.

The world's research scientists and engineers total about 2.5 million. Of these, about 500,000 work only on military Research and Development (R & D). If only research physicists and engineers, those at the forefront of technological innovation, are included, over 50 per cent are working for the military, improving existing weapons and developing new ones. Funds given to military R & D are currently running worldwide at about $80 billion a year.

The skills now monopolised by military science could be rapidly diverted to solve many of the global problems we face. Unless they are these problems will simply get worse. The quality of life for most of the world's people, both in the developed and in the developing countries, will steadily deteriorate.

We will only escape this fate if political leaders emerge who are strong enough to overcome the political pressures exerted by the military–industrial complex and make sure that the resources now devoted to military science are diverted to the solution of the most serious global problems. Given this competition for the use of scientific and technological resources, is the use of such resources to develop non-lethal weapons justified? Or should resources now used for the development of lethal weapons be diverted to the development of non-lethal weapons?

Enthusiasts for non-lethal weapons claim that they will be essential in tomorrow's world for troops fighting asymmetrical wars, and for police and paramilitary forces trying to control the increasing levels of urbane violence in the industrialised and, particularly, in the developing world. Critics of non-lethal weapons argue that they are simply another profit-making product of the defence industries.

Given the importance of the future of force in the broad security context, global and national, *Non-lethal Weapons and Legitimizing Forces* is a very important study, essential reading for all interested in security issues, international affairs, strategic studies, and international politics. Politicians, students, journalists, military personnel, and academics will find it particularly useful.

Frank Barnaby

Acknowledgements

Many individuals have shaped the argument of this book and offered helpful comments. In particular though, my thanks to John Buttle, Mike Coleman, Rob Evans, Alistair Hay, Robert Jones, Tzipi Kahana, Nick Lee, Les Levidow, Nick Lewer, Dominique Loye, Jorma Jussila, Brian Martin, Callum McRay, Birgitte Munch, David Power, Paul Rosen, David Skinner, Brian Woods and Steve Woolgar. Jürgen Altmann made extensive editorial and substantive comments throughout the text. I am grateful to Klaus Landsman for initially setting me down a road that lead to my interest in this topic. Finally, my thanks to Andrew Humphrys and Louise Hulks at Frank Cass, as well as George Pitcher, for their editorial assistance.

In addition, several organizations are worth singling out. Parts of this book have been reprinted by permission of Sage Publications Ltd from Brian Rappert, 2001, 'The Distribution and the Resolution of the Ambiguities of Technology', *Social Studies of Science*, August: 557–92. The Freedom of Information Office of the Civil Rights Division of the US Department of Justice provided many of the documents cited in Chapter 9. The members of the Arms and Security Trade working group of Amnesty International (United Kingdom) deserve thanks for taking in a young student and showing him the importance of concerted collective action in the face of continuing denials by governments of their basic responsibilities. I am grateful to Andrew Webster at the Science and Technology Studies Unit at the University of York (UK) for his years of support to initially pursue the study of non-lethal weapons, as well as the School of Sociology and Social Policy at the University of Nottingham (UK) for enabling me to undertake the writing of this book. The members of the Omega Foundation, particularly Robin Ballantyne, supplied many of the photographs for the book and have been a continuing source of suggestions and inspiration.

Abbreviations

ACPO	Association of Chief Police Officers
APM	Anti-Personnel Mines
BTWC	Biological and Toxin Weapons Convention
BZ	3-quinuclidinyl Benzilate
CAJ	Committee on the Administration of Justice
CBDE	Chemical and Biological Defence Establishment
CN	Chloroacetophenone
COT	Committee on Toxicity [UK]
CR	o-Chlorobenzylidene Malononitrile
DERA	Defence Evaluation and Research Agency
DSAC	Defence Scientific Advisory Council (UK)
EMP	Electromagnetic Pulse
FBI	Federal Bureau of Investigation
HEAP	Human Effect Advisory Panel
ICRC	International Committee of the Red Cross
JNLW	Joint Non-Lethal Weapons (Program or Directorate)
MC	Methylene Chloride
MIBK	Methyl Iso-butyl Ketone
NATO	North Atlantic Treaty Organization
NIJ	National Institute of Justice
NLW	Non-Lethal Weapons
NGO	Non-Governmental Organization
NTAR	Non-Lethal Technology and Academic Research
OC	Oleoresin Capsicum
PAVA	Pelargonic Acid Vanillylamide
PSDB	Police Scientific Development Branch [UK]
R&D	Research and Devleopment
RoE	Rules of Engagement
RUC	Royal Ulster Constabulary
SIPRI	Stockholm International Peace Research Institute
STOA	Scientific and Technical Options Assessment [of the European Parliament]
US	United States of America
UK	United Kingdom
WTO	World Trade Organization

1

Introduction

The use of force is a moral, social and medical problem. When members of the police, military, paramilitary, non-state groups or individuals apply force, questions need to be asked as to whether it was effective, necessary and proportional. Evaluations of such matters are often tied up with the effects of particular forms of weaponry and equipment. Consider the following account of the operation of United States (US) forces in Somalia and the choices made about the appropriate force response:

> In early 1995, some U.S. Marines were supplied with so-called dazzling lasers. The idea was to inflict as little harm as possible if Somalis turned hostile. But the Marines' commander then decided that the lasers should be 'de-tuned' to prevent the chance of their blinding citizens. With their intensity thus diminished, they could be used only for designating or illuminating targets.
>
> On March 1, 1995, commandos of U.S. Navy SEAL Team 5 were positioned at the south end of Mogadishu airport. At 7 a.m., a technician from the Air Force's Phillips Laboratory, developer of the lasers, used one to illuminate a Somali man armed with a rocket-propelled grenade. A SEAL sniper shot and killed the Somali. There was no question the Somali was aiming at the SEALs. But the decision not to use the laser to dazzle or temporarily blind the man irks some of the nonlethal-team members. 'We are not allowed to disable these guys because that was considered inhumane', said one. 'Putting a bullet in their head is somehow more humane?'[1]

During the same year, much deliberation had taken place in an international forum regarding similar technology. State Parties of the United Nation's Convention on Certain Conventional Weapons discussed the possibility of prohibiting blinding lasers.[2] In previous years, organizations such as Human Rights Watch and the International Committee of the Red Cross (ICRC), along with some governments, had pressed for controls on the use of these weapons. Against the initial opposition from the United States and the United Kingdom, advocates of restrictions argued that blinding lasers would cause superfluous injury and unnecessary suffering and thus be prohibited under international humanitarian law. In part, the logic given for this was somewhat counter-intuitive. Humanitarian groups argued that the casualties incurred during war by conventional weapons were not as severe, frequent or long-term as might be assumed. Whatever the intent of soldiers on the ground or military commanders in bunkers, the lethality of warfare with conventional arms should not be taken for granted. In contrast, the deployment of tools designed to blind opened up the potential for permanent damage that would almost certainly bring severe physical, psychological and financial consequences. However horrific the loss of a limb or the reduction in mobility caused by shrapnel or bullet wounds, no prosthesis exists to aid those who cannot see. The question which needed to be addressed, then, was not whether it was better to be dead or blind, but whether the almost certainty of blinding was more inhumane than the possibility of injury or death through conventional arms.

Such arguments followed from various experts' meetings in the late 1980s and early 1990s held by the ICRC to determine the threat posed by laser technology. Silent, portable, low cost and effective, blinding lasers were presented as a step too far. The possible use of these weapons outside of war, say in terrorists' attacks on civilian populations, was so disturbing that governments had to act before the technology became diffused. In the end, calls for restraint prevailed. In September 1995, the US Department of Defense announced a ban on development of lasers specifically designed to cause blinding to unenhanced vision. In October of the same year, State Parties of the Convention on Certain Conventional Weapons adopted a new protocol that placed limits on blinding lasers.

The quote about experiences in Somalia asks the reader to consider the appropriateness of general restrictions on weapons in relation to an alternative context where the class of technology in question is presented in a rather different light. Here, a form of merely dazzling non-lethal weaponry offers a more humane force option. The account is premised on a number of assumptions about the intent of the users of weapons, the likely operation and effects of the dazzlers in specific situations, and the necessity of their use therein. The possible development of dazzling weapons raises a number of questions: How can dazzling and blinding lasers be differentiated? What would the deployment of the former mean for attempts to prohibit the latter? In the chaotic situations of conflict, how likely are dazzlers to inadvertently blind civilians and others? Who is competent to judge such matters? Various questions could be asked about the merits of non-lethal weapons in specific examples. In the case above, could the SEALs have ascertained whether the Somali was 'temporarily' blinded before needing to resort to deadly force? What would have happened until and when the temporary effects ended? Thus, the manner in which the legitimacy of particular force options is decided is a matter of concern and much potential debate.

This book addresses the legitimacy of weaponry in conflict situations. It examines past, present and future intersections of expertise, politics, science and technology through a consideration of the growing attention paid to non-lethal weapons in military, police and incarceration circles. But what does it mean to speak of force as having legitimacy? Beetham has suggested that actions are legitimate to the extent that they conform to established rules, that those rules can be justified by reference to beliefs shared by both the dominant and the subordinate social groups, and that there is evidence of consent by the subordinate to the particular power relation.[3] This would require, for instance, that force be applied in accordance with the limits of the law and that these rules only sanctioned force that was necessary and proportionate. Such determinations, in turn, depend on assessments of the likely effects. Moreover, though, new weapons alter standards for what counts as proportionate and appropriate responses by helping to define the range of possible actions.

While the meaning and appropriateness of the term 'non-lethal' is much contested, as a starting point, these are weapons that are supposed to minimize or at least reduce the severity of injury. In doing so, they are meant to entail comparatively more acceptable options than other force means. Today there are a variety of non-lethals in the arsenal of Western police and military forces, and research and development continues apace into new ones. Among the current options include kinetic projectile munitions such as plastic bullets; chemical irritants such as tear gas; and electroshock devices such as stun guns. To believe some, these 'old-fashioned' technologies are soon to give way to a far broader spectrum of possibilities: acoustic weapons which shatter windows and cause internal damage; electromagnetic pulse beams designed to knock individuals down and cause seizures; and chemical agents which act as calmatives. Add to these old and new weapons innovations in the delivery systems and the combination of different types of non-lethals (for instance, projectiles which release chemical irritants on impact), and it is possible to imagine something of the diversity of potential force options. In recent years, this form of weaponry figured prominently in the highly publicized policing of public protests in Seattle and Genoa, though their possible areas of deployment extend well beyond such instances to include routine policing, incarceration and military interventions.

Despite attempts to foster a positive image of this class of weapons by proponents, their merits are disputed. The use of force by security apparatuses of the state and others is often contentious and the utilization of particular weapons in such acts has been a source of some unease. From the hills of Los Angeles to the streets of Nairobi, concerns have been voiced. Here, the potential for abuse with non-lethals, their less than harmless effects, their unintentional consequences, and their potential to be used as a complement to 'lethal' force makes them less than benign.

This book considers how the promise of technology is created, sustained and questioned. I wish to illustrate how a detailed examination of non-lethal weaponry requires addressing a host of issues about the functioning of expertise, the relations between citizens and organizations of the state, the trust held in modern institutions and the regulation of technology. In these matters, considerations of power and authority loom large. The reader is asked to view accounts of non-lethal weapons as attempts to both inform and persuade audiences about the moral standing of technology and those who utilize it. Non-lethal weapons are bound up with efforts to negotiate the respectability of the use of force, where the standards by which such acts are judged are themselves in flux. Discussions about the merits of non-lethal weapons are often characterized by uncertainty over the facts of the matter, where 'technical' issues mixed quite readily with 'ethical' appraisals. Disagreements about the merits of such weapons hinge on interpretations and judgements. Through an examination of the claims and counterclaims about this form of weaponry, this book will illustrate how the technology functions as a site of political negotiation.

SOME INITIAL POSTURING

This book takes as its starting point for analysis the ambivalence of non-lethal weapons. It is generally acknowledged that any technology can be used for good or bad purposes. The suppression of peaceful political dissent is likely to be viewed quite unfavorably by most compared to, say, attempts to defend police officers from belligerents intent on violence. In between these extremes, though, there are vexing questions about what constitutes acceptable force and how such evaluations can be made. Instances of the use of force are occasions for multiple interpretations of what is happening and why. Just whether someone should be labeled a rioter, a potential security threat, a peaceful protester or an innocent bystander is often contested.

Still, for some individuals, any recourse to weapons is regrettable. The idea of a non-lethal weapon might appear a contradiction in terms if not an anathema. Weapons are, by definition, designed or used to inflict injury and thereby possibly cause death. Yet, as long as one maintains that the resort to some form of force is necessary in some situations, questions must be asked about how violent that force should be. A prohibition against all non-lethal weapons presumably would include devices such as the police baton. Even the harshest critic is likely to acknowledge that non-lethal weapons could have advantages in some circumstances. In the United States, the phrase 'suicide by cop' has emerged to describe circumstances when individuals with weapons confront the police in an effort to provoke them to fire. In such situations, there are pressing questions about the most appropriate response.

Much of the ambivalence of 'non-lethal weapons' derives from the difficulty of classifying the disorderly and diverse set of practices and technologies associated with the term. Classification is itself an attempt to suppress messiness of the world and provide a basis for comprehension and evaluation.[4] There are many ambiguities associated with this class of weapons that make moral determinations problematic. What counts as a non-lethal weapon is unclear. By definition, a weapon is intended to inflict damage and therefore has the potential to kill, so categorizing a particular weapon as lethal or non-lethal would be difficult even if there were consensus on the nature of the potential risks entailed. As will be illustrated in this book, though, such a consensus cannot be taken for granted. Related questions arise in disputes about where to cast blame when things go wrong, whether that be with the technology, its user, the intended target or the volatility of the situations in question. Still further questions can be asked about the right criteria for assessing non-lethals; whether that is merely in terms of the direct physical effects or whether other factors, such as the possible consequences for the escalation of violence, ought be considered. In short, there are considerable evaluation and classification problems.

In starting with a recognition of the ambivalence of these weapons, I do not wish to simply suggest that there are different views about their acceptability and appropriateness. While this is apparent, it is not necessarily a helpful observation. Rather, in drawing attention to matters of ambivalence and ambiguity, I want to set the analysis of this book on a particular footing. Ambivalence is not treated as an innate property of these technologies or situations; rather, it is the outcome of social and organizational processes. Accounts about this class of weapons are attempts to establish ethical relations between technology and humans.

How this is done is a matter that requires attention. As suggested in the case of dazzling and blinding lasers, any assessment of non-lethal force begs questions about what options are being compared, by what criteria, in relation to what circumstances and by whom. Complex technical, legal, medical and operational factors can enter into calculations regarding the appropriateness and acceptability of the use of force. The ambivalence of non-lethals raises a number of questions regarding how statements about them are framed and the strategies by which such framings are advanced.

Moreover, the ambivalence of non-lethal weapons matters in the way in which questions about responsibility, praise and blame are negotiated. When a technology with uncertain or disputed effects, whose proper use requires following highly prescriptive rules, is introduced into a volatile setting, then there are important questions about how disputes about the appropriateness of force are managed. Cavalier attempts to simply adjudicate on the merits of weapons once and for all are bound to be open to question.

This orientation contrasts with attempts to advance a definite reading of technology, in terms of, say, its 'real' effects and whether a particular weapon is actually lethal or non-lethal. While such appraisals are obviously important and necessary, this book seeks to step back from specific claims to consider their basis and implications. This is important because installments of the latest technical or operational information about particular weapons alone often offer little hope of clarifying, let alone resolving, disputes. There are important issues at stake regarding how we discuss technology and how debates are conducted.

In this book, I appraise the promise and threat posed by non-lethal weapons in terms of both how such technologies are instruments of control for acting in the world, and how characterizations of them are actively managed. I argue that these weapons need to be understood in terms of both of these two fundamentally intertwined aspects. Without a careful examination of how accounts of these devices are presented and handled, the analysis risks taking for granted issues that might otherwise be questioned.

The contingencies and assumptions entailed with how notions about the legitimacy of non-lethal weapons are advanced will be put under close scrutiny in this book. I consider how organizations advance particular types of arguments and the justifications for such positions. In doing so, my role is not simply one of apologist for the technology, opposed critic to it or technician merely trying to clarify the issues at stake without taking any position on the relative merits of arguments.

It is not the intention of this analysis to try an authoritative adjudication of contentious issues. I do not want to blind readers with a rendition of the 'real' facts of the matter about a wide range of issues if those could be determined authoritatively. Rather, I intend to acknowledge the contingencies and negotiations at work in any assessment. I advance particular assessments of the merits of non-lethal weapons, while also trying to acknowledge more fundamental issues at stake about uncertainty, disagreement and controversy. In surveying the recent increased prominence of non-lethal weapons, I give a skeptical orientation to claims about individual devices and the overall merits of the class of technology in general: 'skeptical' in the sense that it is prudent to unpack the range of claims and counterclaims made.

Adopting this position, I suggest that taking the uncertainties and disputes associated with this technology as a topic of analysis, rather than trying to offer a definitive or resolved account of weapons, can provide a number of insights. As will be argued more fully in later chapters, in assessments of weapons, technology in general or any matter, there are various ways in which implicit and explicit framings are given to the topic in hand. Throughout, questions are raised about what evidence and inferences would be required to assert particular assessments of non-lethal weapons. In this way, the book attempts to offer a persuasive account of non-lethals while trying to persuade the reader about how to approach the study of technology.

RATIONALE FOR APPROACH

To date, non-lethal weapons have been of limited concern to researchers outside of those agencies concerned with deploying them, and those groups and others that question them. Of the academic work that does exist, much of it stems from peace and security studies, where the main interest is in peacekeeping, or more broadly, applications of this technology in warfare.[5] To the extent criminologists have examined this weaponry, they have been preoccupied with meeting the managerial agenda of their funders in asking questions about the effectiveness of different devices. Recently, legal scholars have begun to ask how the introduction of non-lethals might affect or be controlled by existing international and national laws.[6]

In this book, a diverse range of technologies on offer under the banner of 'non-lethal weaponry' are examined from their initial justifications to their operational deployment. In doing so, a variety of issues will be touched on that are themselves worthy of a far more detailed consideration than is on offer here. Topics such as the weapons acquisition in the military,[7] the use of scientific standards for judging safety,[8] policing cultures surrounding the use of force, and the presentation of accounts of conflict in the media have been the topic of many books and could be the subject of many more. Furthermore, particular instances of deployment, such as political protests or conflicts, could be given detailed consideration.

The purpose here is to provide a window into a diverse set of issues. That window is framed by examining debates about non-lethals and following the arguments where they lead. This broad analysis can be justified on a number of grounds. As will be argued in later chapters, developing and utilizing particular weapons with some degree of legitimacy requires the creation of audiences receptive to such initiatives. In considering the growing importance of non-lethals it is prudent to look at not only specific innovations but also the wider promises on offer with this class of weapons. Commenting on the 'successfulness' or 'unsuccessfulness' of the introduction of a particular device, and the meaning of these terms, requires an institutional as well as technical analysis sensitive to both general and specific promises made.

This broad analysis also enables connections to be made between various stages in the development process that might otherwise be missed. In doing so, it is able to consider how claims about technologies made in one setting get translated into others. For instance, only examining the operational use of a technology would fail to notice how the initial justifications offered change and how the scope for deployments expands or contracts over time as an innovation matures. A narrow analysis also risks glossing over important questions about the functioning and interrelation of expertise. Toxicologists with a scientific training often look into likely health effects; security personnel might be asked to comment on the practicality of the technology in specific situations; criminologists and others might be commissioned to assess the impacts of devices on issues such as the number of assaults to officers; and clinicians might document the nature of injuries and deaths incurred. While it is necessary for experts to work in limited domains of competence, the argument presented here brings together a variety of claims made about non-lethals in order to assess them in relation to one another.

In doing so, however, the obvious risk exists of making poorly thought-out claims across a number of ill-understood areas. I am not a soldier, a toxicologist, a police chief or a lawyer, but rather a social scientist with a fairly broad range of interests within that domain. The situation calls for some caution, but there is much insight to be gained from venturing outside of well-trodden ground. It is a matter for the reader to decide if this has been done competently and to venture outside of the argument presented and make further assessments about its relevance.

As a book that takes as its topic a wide range of concerns about a class of technology, there is also the danger that the argument will be rather thin in places. The chapters ahead mix general overview of technological areas (Parts I and II) with more detailed case studies (Part III). The empirical basis for this book could always be more thorough, and the reader will no doubt want further information about the technologies detailed than is provided. Indeed, one of the central contentions I want to advance is that when one starts unpacking disputes about the merits of particular non-lethal weapons, the desire for more and more information than is typically on offer becomes readily apparent. In such a situation, an analysis that evokes a wide range of connections and interpretations offers the potential to be provocative and challenging.

The topic of non-lethal weapons and related concerns about the use of technology in the application of force are not only worthy of study because they break new topical ground, they also press us to re-examine a number of conceptual issues in the social sciences. So, how authority is exercised by a security force tells us a good deal about the relation between the state and its citizens as well as about the relation between states, but these cannot be understood without a consideration of the means of exercising authority. As detailed below, themes in policing, criminology, peace and security studies, politics, technology studies and policy analysis are integrated into the chapters. This book, though, is primarily an attempt to analyze the legitimacy of a class of technology, so I will not spend a great deal of effort engaging in specialized theoretical debates, though this analysis is immersed in them. Those in search of more thorough considerations of some of the themes discussed can follow the references out elsewhere.

Inevitably, though, some issues will be given more attention than others in the chapters that follow. While activities in a number of countries are considered, most of the analysis is focused on developments in the United States and the United Kingdom. In many respects, the United States is the world leader in terms of the state of its technology and volume of commercial interest. Britain has a long history of the use of non-lethals because of its colonial history and more recent conflicts in Northern Ireland. In the case of non-lethals, the United Kingdom also plays a familiar role of bridging between continental Europe and North America. There are other substantive limitations to this book, as well. Anti-personnel applications are given far more attention than anti-material ones (see Chapter 3).

AN OVERVIEW OF THE CHAPTERS

This book is divided into four parts and eleven chapters. The organization is designed to bring an ever-widening range of considerations to bear. The chapters in Part I provide an overview of non-lethals as well as an initial elaboration of how this analysis approaches the study of technology. Chapter 2 begins by asking a seemingly straightforward question regarding non-lethal weapons: what are they? Far from being a clear-cut issue, however, a detailed consideration of this question raises important issues about how we understand the relation between technology and its situations of use. Drawing on insights regarding the relationship between politics and technology developed by sociologists of technology such as Steve Woolgar and Keith Grint, this chapter establishes the importance of examining how claims are made about technology and why particular stories win out over others. This chapter begins to elaborate themes that are taken up in later chapters about context, claims and classifications as well as agency, assumptions and associations.

Following from these considerations, Chapter 3 provides a fairly conventional account of current and proposed non-lethals. Such weapons are said to offer the possibility of disorientating, dazzling, disabling and calming rather than the shattering of bones or tearing of flesh. In reciting these accounts, the organizations and funding programs in place to foster non-lethal weapons are surveyed. Following on from points raised in Chapter 2, this chapter draws attention to the commitments of particular types of descriptions, the role of interpretation and the bounds of debate.

Chapter 4 outlines the assertions made regarding the range of wide-scale environmental changes which are thought to require additional non-lethal responses: the growing importance of peacekeeping, the changing character of warfare, the continuing threats to domestic order, and the growing unacceptability of the use of force. This chapter considers how such claims are marshaled to justify the development and deployment of non-lethals. It describes the multiple promises accorded to non-lethals at the strategic and tactical levels. By way of contrast, alternative schools of thought from those associated with advocates are discussed. Critics of the technology turn many of the statements of proponents on their head. Divergent interpretations derive from fundamental differences in the framing given to situations and the inferences made by commentators. Through noting the importance of such issues, this chapter highlights the need to interrogate the basis of evaluations and the limitations therein.

Building on the earlier discussion, the chapters of Part II investigate the main areas of discussion about non-lethal weaponry: their effects, their operational use and their control. Appeals to science are often evoked to establish authoritative determinations about the relative harm caused by these devices. Through a series of studies of existing non-lethals, Chapter 5 outlines claims and counterclaims made about the safety of specific weapons and how notions about their effects have been used to justify or criticize their deployment. It is argued that even if one takes well-established technology, like chemical irritants or kinetic energy munitions, it is possible to raise a host of concerns about their safety. In general, what can be said is that the existing publicly available evidence about the safety of particular devices is often weak or mixed and not able to resolve the 'facts of the matter' in an authoritative fashion. The appraisal of weapons is bound up with processes for defining, identifying and interpreting risks that are conditional and speculative.

Whatever lab tests about safety indicate, the appropriate use of non-lethals requires following certain rules that define standards of acceptable force. Chapter 6 discusses a few contested operational deployments of non-lethal weapons in the United States, Israel and the Occupied Territories and Northern Ireland. In doing so, it considers how controversies about particular situations unfold and what implications these have for questions about the legitimacy and proper control of the technology. The possibility of lack of adherence to rules in the use of force is given some attention. As will become apparent, even when there is agreement about unfortunate injuries and 'deviations' from protocols, the actions that should follow are hotly disputed. In such disputes, the importance of context is multi-faceted and varying definitions of it are drawn upon to make sense of particular interpretations.

Chapter 7 then follows on from the other chapters in Part II to consider attempts to govern and restrict the proliferation and development of non-lethals. International treaties and obligations are examined as they relate to the control of different types of weapon. The application of general principles and proscriptions about the proper conduct of conflict to specific uses of non-lethals raises a basic problem: when is it justifiable to cut through complex assemblages of individuals and technologies in order to offer definitive assessments about what prohibitions ought to be in place? Various themes are raised in relation to this problem, such as the policing of controls, appraisals of trust and the negotiation of classifications.

Parts I and II identify a number of key issues and tensions associated with non-lethal weapons: these are then elaborated further through the use of case studies that cover military, police and prison situations in Part III. Chapter 8 evaluates the deployment of incapacitant sprays for routine policing in Britain. This topic provides something of a 'best case' scenario for the potential of non-lethal weapons. As opposed to highly charged public order events or military interventions, routine policing does not evoke the same degree of controversy. In addition, the supposed 'consensual' nature of policing in Britain and the previous experiences with sprays outside of Britain should have meant that their introduction was fairly unproblematic. This chapter examines the dynamics surrounding the introduction and deployment of the devices. In doing so, significant doubts are raised about the robustness of the precautions taken and the wisdom of decisions made. More generally though, this chapter illustrates how the uncertainties associated with the sprays were handled and how these bore on determinations of who was responsible for ensuring their appropriate use at an operational level. In doing so, many of the core conceptual themes discussed in the book are drawn upon.

Chapter 9 analyses a recent major non-lethals initiative in the United States: the introduction of electroshock stun guns and forms of pepper spray to all detention officers in Maricopa County, Arizona jails. This was the first such trial in the United States and has been billed by some as a major success story. Yet, upon closer examination, the merits of the trial are far less certain. Various official and unofficial evaluations were undertaken that made conflicting assessments about the trial and how they ought to be appraised. In examining this, Chapter 9 considers how the assessment of technology can be approached when the facts are disputed and uncertain. A major theme, and one that follows from previous chapters, is the importance of recognizing the interrelation between 'technology' and its 'context' in the production of credible claims.

Chapter 10 takes as its topic the possibilities for existing and future non-lethals in humanitarian interventions, particularly in relation to changes in the overall strategic level of warfare. A variety of prominent theoretical scenarios are described and analyzed, including those by the US Marines and the US Council of Foreign Relations. The assumptions beyond these are interrogated in light of past discussions in the book. Following from this, the recent NATO intervention into Kosovo is drawn on as an example for thinking through the possible strategic and tactical implications of non-lethals. Among proponents of non-lethal weapons, Kosovo is often presented as a lost opportunity to introduce a new form of warfare. Various aspects surrounding the public-relations handling of this intervention are examined so as to consider what implication non-lethals are likely to have on the character and legitimacy of similar conflicts in the future. Drawing on previous arguments, Chapter 11 concludes with a summary discussion and proposes various policy recommendations.

INTRODUCTION

NOTES

1. Pasternak, D., 'The Pentagon's Quest for Nonlethal Arms is Amazing. But is it Smart?' *US News and World Report*, 7 July 1997.
2. Doswald-Beck, L., *Blinding Laser Weapons*, Papers in the Theory and Practice of Human Rights, Number 14 (Colchester: University of Essex, 1995).
3. Beetham, D., *The Legitimation of Power* (London: Macmillan, 1991), 16.
4. Zygmunt, B., *Modernity and Ambivalence* (Cambridge: Polity, 1991).
5. Dando, M., *A New Form of Warfare* (London: Brassey's, 1996) and Lewer, N. and Schofield, S., *Non-Lethal Weapons* (London: Zed, 1997).
6. Fidler, D., '"Non-lethal" Weapons and International Law', *Medicine, Conflict and Law*, 17, 3 (2001), 194–226 and Krüger-Sprengel, F., 'Impact of Non-lethal Weapons on the Law of Armed Conflict and Disarmament', in *Non-lethal Weapons: New Options Facing the Future*, Proceedings of the 1st European Symposium on Non-lethal Weapons, 25–26 September 2001, Ettlingen, Germany (Postfach: ICT, 2001).
7. Farrell, T., *Weapons Without a Cause* (Basingstoke: Macmillan, 1996).
8. See, for instance, Kammen, D. and Hassenzahl, D., *Should We Risk It?* (Princeton, NJ: Princeton University Press, 1999); Klapp, M., *Bargaining with Uncertainty* (New York: Auburn House, 1992); Morone, J. and Woodhouse, E., *Averting Catastrophe* (London: University of California Press, 1996); and Thornton, J., *Pandora's Poison* (Cambridge, MA and London: MIT Press, 2000).

PART I:
CLAIMS AND EXPECTATIONS

2

What's in a Name?

This chapter begins the detailed examination of non-lethal weapons by addressing a basic issue: what are they? The answer given to this question has obvious importance for analysts in defining their scope of study as well as for practitioners in attempting to determine the potential of this category of weapons. Although that is seemingly a straightforward matter with a presumably straightforward answer, in this chapter I explore why this is not the case.

The purpose of this chapter is two-fold. First, it examines the definition of non-lethal weapons. Much debate has already taken place as to whether the term itself is misleading by implying that no fatalities result from these devices. The focus here is not so much on this issue, but on the more general question of what differentiates lethal and non-lethal weapons. In addressing this, I shall raise a variety of points about how we approach the consequences of weapons and what knowledge is important in assessing them. A consideration of the problems associated with attempts to categorize and assess non-lethal weapons leads to a much more basic question about the relation between technology, its context and our knowledge of both. By considering such issues, the second purpose in the chapter is to elaborate the theoretical approach adopted in this book.

MORE THAN WORDS

The desire for weapons with effects other than death is not new. Military and police forces have long sought technologies and tactics that enable a differential application of force, sometimes with a view to causing limited harm. One of the oldest known instances of such devices is the dispersal of pepper and oil by ancient Chinese armies to irritate and temporarily incapacitate their opponents. The police truncheon can be regarded as a simple form of non-lethal weaponry that has been around for some time. Although it can kill, it is seldom used for this purpose and death rarely results in practice. Yet the term 'non-lethal' itself has only gained widespread currency in military and police circles during and since the 1990s. At an initial level, the term is meant to contrast particular weapons with standard (read: lethal) ones. For proponents, their development is portrayed as enabling a wider range of force options and thus more appropriate responses.[1] Just how significant and how progressive the possibilities opened up are matters of some disagreement, even among the converted. The writers Alvin and Heidi Toffler have gone so far as to suggest that a new arms race has started to find weapons designed to keep people alive;[2] a characterization likely to raise more than a few eyebrows.

Most analyses of this class of non-lethals begin with a description of the instruments that fall under this heading. I want to resist the temptation to begin in this way, in order to step back from technical descriptions and instead consider what is at stake in the label of 'non-lethal'. As will be argued, it is not just that non-lethals are difficult to define, but that alternative definitions point to fundamentally different ways of making sense of the contribution of non-lethals to conflicts. The assumptions implicit in the definition of non-lethal weapons inform notions of their legitimate deployment and thus the proper standards for their control.

Initially, we can begin by noting that the appropriateness of the expression 'non-lethal' has been a topic of much dispute since the early 1990s, when members of the US military began actively promoting the term.[3] The basic counter has been that no weapon is free from the risk of fatalities. A repeated joke is that the perfect non-lethal weapon would be marshmallows descending by parachute.[4] On the other hand, though, enough marshmallows will kill you if placed properly.[5] At best, for some, then, the term 'non-lethal' is a misnomer. At worst it is an outright deception. Describing a weapon as such, as opposed to being lethal, might give it greater public-relations credibility, but this description is not necessarily accurate in practice.[6] In the Gulf War, for instance, tiny carbon fiber spools caused short circuits in power plants and switching stations in Iraq. The likely resulting deaths from the lack of electric supplies cannot be understood as non-lethal in character.[7]

Given such evaluations, it is not surprising that alternative phrases such as 'less-lethal', 'less-than-lethal', 'pre-lethal', 'worse-than-lethal' and 'soft-kill' have been advanced to designate weapons others would prefer to call non-lethal. In many law enforcement agencies in the United States and Europe, less-lethal is widespread because of the possible liability repercussions that may follow from deaths incurred by weapons advertised as fatality-free.[8] 'Pre-lethal' points to the potential of weapons to enhance the lethality of others.[9] The Omega Foundation, for instance, has collected hundreds of accounts of non-lethals being used in conjunction with, sometimes prior to, firearms.[10] The term worse-than-lethal highlights the probability for injuries such as blinding which, as discussed in the last chapter, have been argued to be more severe than those incurred by weapons in conventional warfare. 'Soft-kill' refers to the targeting of entire areas, causing effects to both individuals and technologies by the same form of activity so as to render them incapable of fighting, which it is sometimes argued could in fact entail death.

For some proponents, the problems associated with various names are much to do about nothing. John Alexander, former director of non-lethals research at the US Los Alamos Laboratory, for instance, argues the dispute about the right terminology is a matter of mere semantics.[11] The term non-lethal, while not perfect, expresses the intent to minimize or at least reduce casualties. And yet, as is clear by the designation of pre-lethal and worse-than-lethal, there is some discussion about whether this intent is indeed the prime motive. As will be discussed more thoroughly later, alternative terms derive from different assessments made of the intent of users, weapons' effects in practice, the situations of use and the appropriateness of targeting particular individuals. A 2000 joint US/UK Non-Lethal Weapon Urban Operation Wargaming Program concluded that the term non-lethal was 'inaccurate' and might lead to confusion for the media and the public, but pragmatically concluded that 'the official definition will be irrelevant since it cannot be mentioned or explained in a 30-second sound bite'.[12] Yet as this book does not seek to package its claims for 30-second sound bites, it might be worth spending more time examining the issues at stake.

After some years of vacillation between non- and less-lethal terminology, non-lethal has become the preferred term by those in the US military and policy circles. As much of the worldwide impetus for this technology derives from the United States, this term has taken hold in many international discussions. The US Department of Defense defines non-lethal weapons as 'discriminate weapons that are explicitly designed and employed so as to incapacitate personnel or material, while minimizing fatalities and undesired damage to property and environment'.[13]

Even taking this term and definition as being imperfect but reasonable descriptions does not resolve disputes about what should count as a non-lethal weapon. The literature about non-lethals includes a wide range of weapons, not all of which figure into every analysis. So long as one accepts the term, while there are a range of technologies such as truncheons or tear gas that are generally seen as non-lethal, there are those with arguably more tenuous links. For some, the disruption techniques of information warfare (such as computer viruses), as well as the persuasion techniques of psychological operations (such as the distribution of pamphlets), fall under the non-lethal heading.[14] With such a broad definition comes the obvious potential of conflating a variety of technologies and activities with varied purposes. Long-time proponents of non-lethal weapons Janet and Chris Morris have gone as far as to suggest that airpower itself might be regarded as a non-lethal weapon because of the possibility of targeting objects discriminately through the use of modern guidance systems.[15]

Any determinations of what constitutes a non-lethal weapon hinges on assumptions about the manner in which the technology is used in practice. Physicist and long-time security commentator Edward Teller is reported as suggesting 'mini-nukes' could function as non-lethal weapons so long as civilian populations could be evacuated from affected areas in time.[16] In this case, to the extent that the status 'non-lethal' confers a greater legitimacy on the use of particular weapons, this depends on evaluations concerned with the direct and immediate effect of weapons in relation to a particular set of operational conditions.

Designating a weapon as non-lethal because of the possibilities it affords for a discriminate response, and thereby in some way viewing it as preferable to a lethal weapon, ties in with broader debates about the legitimacy of force. Whether it is said to be reasonable and appropriate often depends on evaluations made of what a technology can do, how it is said to function in practice and whether such options are acceptable in given situations. During autumn 2000, as conflict heightened between Israelis and Palestinians, the Israeli Air Force used helicopter gunships in attacks in the West Bank and Gaza strip. Such technology was said to allow the pin-point targeting of key strategic facilities that minimized potential casualties, particularly as warnings were given, in at least some situations, of attacks. Human rights groups, such as Amnesty International, in turn made counter-assertions about the injuries and deaths caused in particular attacks, the less than discriminate targeting at work and the failure to give adequate warnings.[17] The excessive use of force entitled by the use of gun ships against civilians and the Palestinian Authority escalated the level of violence. Amnesty appealed to governments, such as those of the United States and Britain, to suspend exports of supplies for weapons used to commit human-rights violations.

THE CALCULUS OF LIFE AND DEATH

The points made in the paragraphs above suggest that distinctions made between weapons deserve close scrutiny. What bases have been offered for differentiating lethal and non-lethal weapons? How do criteria for assessing technologies in terms of their lethality enter into evaluations? In what way are the distinctive characteristics of non-lethals said to confer a greater legitimacy to the use of force? In other words, what makes a non-lethal weapon 'non-lethal'?

It seems reasonable to assume that lethal and non-lethal devices can be differentiated by comparing their effects. While nearly any weapon (or object, for that matter[18]) can be used to cause serious injury or death, non-lethals are supposed to reduce the chances of such outcomes. In an illustration of the differences between non-lethal weapons, John Alexander provides a side-by-side comparison of kinetic-energy weapons by firing them into gelatin blocks.[19] In one instance, the penetration depth of a beanbag (a small metal shot with a canvas bag wrapped around it) is compared to that of a 9mm pistol round. While the beanbag enters two inches into the block, the pistol round enters fourteen inches. The result is taken to 'unambiguously' demonstrate the beanbag's non-lethal effects as compared to those of a conventional round. Whether or not this test is a realistic demonstration of the merits of the beanbag or non-lethal kinetic-impact weapons in general will be considered in more detail in Chapter 5. For now, two points are worth noting. First, the comparison is premised on the ammunitions being fired in the same situation. Second, even if one were to accept the conclusion that some weapons are less lethal than others in a general sense, what this means in practical terms is less than clear. Questions must still be asked about when recourse to a particular weapon is justified and how assessments of relative effects should enter into decisions about what force options to employ.

There have been some attempts to determine effects in a systematic manner so as to provide a basis for classification. Becker and Heal[20] have proposed employing standard military 'probability-to-kill' measures as a basis for determining lethality. Here, weapons used in a given situation are compared in terms of the likelihood of death resulting. In practice, such classification schemes have failed to be taken up very widely. Not all weapons are amenable to this sort of quantification; probability-to-kill measures have developed out of the study of impact and explosive weapons that impart a certain amount of energy in a given situation. Moreover, specifying effects in a general and abstract way is of limited relevance in understanding the implications of weapons. Such assessments rely on a set of (often unstated) presumptions about how technologies will be used in practice. For instance, a common assumption underlying determinations of the dangerousness of kinetic weapons is that these will not strike sensitive areas such as the head. How often this is the case (depending on a weapon's accuracy or the volatility conditions of its use) affects lethality determinations.

But if concocted experiments or statistical comparisons are of limited help in distinguishing between lethal and non-lethals, neither are aggregated figures based on experience. One of the most systematic studies of wartime casualties, conducted by the US Office of Operations Research, found that over a number of wars in the twentieth century, 10,000–50,000 bullets were fired for every person hit.[21] The International Committee of the Red Cross[22] has collected casualty rates for many recent major wars and found conventional fragmentation weapons and rifle hits 'only' kill roughly 20–25 per cent of the time. Combining such figures suggests a very low kill probability for what would normally be considered lethal weapons. By contrast, evidence gathered on the number of deaths by plastic bullets in Northern Ireland suggests 1 in 18,000 rubber bullets fired resulted in death and one death occurred for every 4,000 plastic bullets fired.[23] A report by the American Civil Liberties Union 'associated' death with 1 in every 600 cases of people sprayed with pepper spray in California between 1993 and 1995 in instances of the incapacitant's use against those restrained in police custody.[24] The counter-intuitive 'lethality' implications raised by these lines of probabilistic argument and the assumptions buried within them all point to the limitations of separating a weapon from its context of use for the purposes of determining its abstract lethality.

In alternative situations, a given weapon will be employed in alternative ways and the possibility of death resulting will differ significantly. So, despite the fairly widespread use by special tactical units in the United States of kinetic weapons similar to the plastic bullets deployed in Northern Ireland, there have been fewer fatalities associated with the former.[25]

Complicating the picture still further is that many of the claims about deaths incurred by particular weapons are disputed. In the case of Northern Ireland, for instance, Chief Superintendent Colin Burrows argues that the generic term of 'plastic bullet' does not do justice to the types of modification that have taken place in the design of the bullets and their launcher, as well as training improvements, in the last 25 years.[26] Demonstrations or stand-offs, as in the events surrounding the 1996 marching session, saw thousands of plastic bullets fired without any fatalities. Likewise, the casual link between in-custody deaths and pepper sprays in the United States has been the topic of much debate.[27] These examples are discussed in more detail in Chapter 5. Sufficient to note for now that in many cases there is no firm agreed empirical basis for simply comparing fatalities or injuries.

Where the limitations of effects-based approaches for distinguishing between weapons become most apparent is in decisions about the necessity of controls. In relation to conventional anti-personnel weapons, Prokosch[28] recounts a number of recent international arms-control negotiations. The contingency of the consequences of technology has not gone unappreciated in such discussions. What is different in these negotiations compared to the examples given above is that the contingencies entailed with effects have provided a platform for those wishing to downplay the severity of weapons. Various lines of counter-argument have been made against attempts to establish definitive (and condemning) classifications of weapons such as cluster bombs and incendiary devices: Did critics really understand the technology? Was it really so 'all or nothing' in its impact? Aren't the effects largely dependent on the intelligence information obtained and the precision of targeting mechanisms? Endless debates over the correct calculus for determining death and injury have stifled the negotiation of treaties designed to place proscriptions on weapons.

MORE THAN EFFECTS

For a host of reasons, even for prominent spokespersons, a strictly effects-based approach has been taken as an insufficient basis for classifying weapons. While common sense suggests that it would be preferable (though hardly desirable) to be shot with a bean bag rather than the average bullet, it is also necessary to be skeptical about how such comparisons are formulated and what inferences should be drawn. Despite some exceptions,[29] in addition to some reference about effects, for proponents what makes non-lethals 'non-lethal' is the *intent* behind this technology. Non-lethals are intended to minimize injury. Military strategists Lovelace and Metz, for instance, have argued that non-lethal weapons can be distinguished from lethal ones in some cases only in that the former were not primarily designed to cause death. In real terms, the damage may be much the same.[30] John Alexander expresses the importance of intent rather bluntly: 'Non-lethal weapons are a goal based on intent.'[31] The US Department of Defense definition of non-lethals as 'discriminate weapons that are explicitly designed and employed so as to incapacitate personnel or material, while minimizing fatalities and undesired damage to property and environment' implicitly gives a central place to intent.

In a matter that also speaks to the limitations of assessing direct physical damage, Lt. Sid Heal of the Los Angeles Sheriff's Department argues that non-lethals differ from conventional ones in two important respects.[32] First, they are designed and employed to defeat the recipient's *will* rather than *ability* to resist.[33] As opposed to conventional weapons that are rated on their damage-inflicting characteristics, non-lethals are meant to target the resolve of recipients. This might entail causing serious damage, but it might not. Second, following on from this point, non-lethal devices are defined by the intent behind their design and deployment rather than by their effects. So, while death might result, this is not an aim.

Arguably, the emphasis given to intent qualifies the limits to harm while also reinforcing the legitimacy of non-lethals. To put it in an extreme form, whatever the actual effects of these weapons in practice, the focus on intent expresses the driving aim to minimize injury and death. Of course, in the 'wrong hands' any technology can be used inappropriately. Abuses associated in non-Western regimes with non-lethals are not surprising for some.[34] US military analyst Robert Bunker has gone as far as to suggest that while Western non-lethal weapons aim to minimize injury, non-Western non-lethals are those intended to cause long-term disablement.[35]

The promise of non-lethal weapons, then, requires forwarding a particular reading of why force is used. How this can be secured is a matter of great concern.[36] Just whose intent is deemed to provide the basis for assessment is part of this process. If the way in which non-lethals are defined is meant to reinforce a view of the progressive intentions of certain parties sponsoring their development, it is also thought by some to absolve responsibility. Here, the status of a weapon as 'non-lethal' ensures a certain moral credibility is built in. For instance, in the mid-1990s, the British-based company Pains Wessex came under criticism from human-rights groups about its export of plastic bullets and tear gas to places where they were used as a form of crowd punishment. Responding to such criticisms and the need for stricter accountability in export controls, a company official is reported as having said, 'What do they want us to be transparent about? We don't make things that kill people.'[37]

One of the contentions of this book is that the manner in which discussions of intent are framed – whether that be in terms of the motives and actions of manufacturers, governments, operators or recipients – plays an important role in assessments. A prevalent manner of discussing intent is in relation to those parties directly involved in particular encounters. Explicitly throughout discussions of this technology, the point is made that these weapons are just tools. Non-lethals are good or bad depending on the way they are used. Like any tools they can be abused, but like any tools they can be used in an ethical way.[38] The emphasis on intent is shared across a wide range of commentators, though from which end of a weapon the matter of intent is considered depends on who is cast as victim or villain. Area acoustic weapons capable of shattering windows and damaging internal organs are said to allow recipients the opportunity to stay or flee a scene of their own volition.[39] Peace researcher Nick Lewer has portrayed the key issue in evaluating non-lethals as one of how controls can be placed on weapons whose effects depend on the intent of their users.[40]

THE IMPORTANCE OF BEING SKEPTICAL

The discussion of non-lethal weapons has so far drawn attention to a number of basic points about this class of technology. To begin with, proponents and detractors generally agree that the categorization of a weapon is relational; this in the sense that whether a weapon is deemed non-lethal is not established by reference to some absolute standard (say zero fatalities). I built on this to further suggest that the attribute of a weapon as non-lethal is not an inherent property, rather it is a matter of the sort of context-dependent interpretations that are made. The lethality of a device has little to do with its physical properties in and of themselves as if these could be understood independently of how the weapon is utilized in practice. The notion of non-lethality is not a concept that exists on its own: it is tied in with concepts such as proportional and necessary force. Understanding the 'in practice' employment of a weapon requires considering how 'technical' issues about their capabilities intermingle with 'social' considerations about how those devices are used and given meaning. This in turn draws attention to what is known about weapons and how it is known.

The discussion so far would suggest the need for care in making statements. Definitive claims about the capacity of non-lethal weapons – statements about what they are, what they do and what they cannot do – are likely to be contestable. This book takes as its platform for understanding technology the work of sociologists Steve Woolgar and Keith Grint.[41] Their approach is one of 'analytical skepticism' toward statements made about technology; skepticism in the sense that it is necessary to unpack the claims and counterclaims made.[42]

While not denying that technologies can constrain or enable action, the key questions for Woolgar and Grint are how claims about constraints or enablers are established. Statements that death or injuries are inevitable in certain situations should beg a number of questions about why this is the case and how such an inevitability is seen to explain or justify certain outcomes. Rather than seeking to determine the actual constraints or enabling characteristics of technology, Grint and Woolgar instead examine how claims are made about these and why particular ones win out. In relation to non-lethal weapons, for instance, it is unlikely in most situations that operators will have extensive experience with the devices. Even in the highest-security prisons in the United States, individuals rarely use such devices.[43] That the use of force is not a regular occurrence means that the way in which people form accounts is central to understanding the range of meanings and attributes given to technologies.

Thus, the telling of accounts about what devices can and cannot do are part of the process of constituting their effects.[44] For example, a source of concern about the deployment of non-lethal weapons is whether they might escalate conflict. Some of those within as well as outside of security forces have contended that the design of weapons as 'non-lethal' encourages greater usage and thus, paradoxically, the need for stricter controls on these devices than on 'lethal' ones.[45] In such a situation, understanding what the introduction of a particular weapon means requires being sensitive to how notions about it are formed and develop over time.

For Woolgar and Grint, an account of the definitive properties of technology that takes certain properties as natural and beyond question 'unnecessarily compromises our ability to challenge the more foundationally ingrained sources of power'.[46] Rather than waiting for the proper authorities to determine the facts, it is instead considered how facts are established. Striving for 'a systematic understanding of which kinds of versions [of technology] win the day, in what circumstances, and why'[47] avoids closing off interpretations because it does not attempt to make definitive claims about the truth of the matter. Rather, the truth of the matter is subject to consideration about how it becomes considered as such. Because of this, appeals to the real properties of technology as a means of arbitrating between claims are somewhat wanting. In analysing technology, Woolgar advises:

> to investigate, rather than accept at face value, the claims descriptions and other representations of technical capacity. Thus analytical scepticism takes it as axiomatic that every representation of technical capacity is situated, motivated, constructed; that it is contingent upon the time and circumstances in which it is produced. This is not, by the way, the same as cynicism. Whereas cynicism dismisses the possibility of describing technology, analytical scepticism turns the contingency to an advantage. It makes it the basis of what it

26

is that a technology is said to do ... [A]nalytic scepticism encourages us always to ask – in the face of any technical description – *who* says that the technology will do this, *how* do they say this, *when*, *where* and *why* is this description given?[48]

To be sure, this general orientation implies a particular way of studying technology. While in our everyday lives most of us get along in making calls by a telephone, arriving to work with some form of transport, or typing on a computer without questioning the devices deployed, in situations where there is disagreement, uncertainty and unknowns, such skepticism has many advantages. A challenge addressed in this book is how the approach advocated by Woolgar and Grint can inform the assessment of non-lethal weapons.

MAKING SENSE OF CLAIMS, CONTEXTS AND TECHNOLOGY

For now, it is worth making a few points about what analytical skepticism implies for the investigation that follows. First of all, the assumptions about the capacity of non-lethal weapons are key in the evaluation made of it. Whether a weapon is actually 'non-lethal' or 'lethal' in its effects and whether non-lethal weapons really offer novel ways to think about conflict resolution, for instance, are central concerns in the justifications or criticisms made. While this analysis needs to be concerned with the answer to such matters, I also want to take a step back from these claims and take them as a topic of analysis. In doing so, certain questions become centre stage: What sorts of claims are on offer? How are particular claims substantiated; what evidence and lines of reasoning are assembled? What functions do these claims play in deliberations about the merits and drawbacks of deploying technologies? In other words, the emphasis here is 'fact-breaking' rather than 'fact-building'; this in the sense that statements made about what non-lethals can and cannot do are traced back toward a consideration of how such 'statements of fact' were produced, rather than just taking statements to justify particular arguments.[49] That much of the existing discussion about non-lethal weapons is highly optimistic means that much of the analysis that follows examines the production of such positive assessments. In practice, of course, some fact-building must be done for the argument here to proceed. But as this is done, close scrutiny needs to be given to how.

A second issue stemming from 'analytical skepticism' concerns how weapons are discussed. In asking who says what about technology, how, when, where and why, it is necessary to be aware of the implications of the characterizations made and the conventions in language. To describe a specific event as a demonstration, a public-order disturbance or a riot suggests a particular interpretation of that event and what types of force intervention might be justified. Labeling non-lethal weapons as 'weapons' is itself contingent and not value free. Many in the police, for instance, refer to chemical incapacitant sprays and plastic bullets (the latter called baton rounds) as personal defense equipment. As will be demonstrated, different designations of events or objects become incorporated into forms of argument in a manner that suggests a particular understanding of the pertinent issues in question. Mere 'descriptions' are not simply that, but form part of the evaluation process.

More fundamentally though, discussions can incorporate a variety of assumptions about the properties and effects of weapons.[50] For instance, as I started into the writing of this book, various questions were being asked in the media about the appropriateness of carpet bombing and the use of cluster bombs by American forces in Afghanistan.[51] The basic question was whether these amounted to unacceptable force options, because they were disproportional or indiscriminate. Phrased in these blunt terms, though, discussions about the merits of these activities are somewhat problematic. 'Carpet bombing' or 'the use of cluster bombs' represent broad activities that can be taken to imply any number of specific features. It is not just that the terms are too broad, but assessments of these actions require drawing on various contextual assumptions. So, where is this bombing taking place, against whom, how effective is it, and how is this information known? In practice, those asked to judge the appropriateness of cluster bombs or carpet bombing do so in relation to background considerations about their likely use (such as to what extent civilians might be killed), the trust held in those institutions who utilize and report about these tactics, as well as determinations about the overall appropriateness of any bombing. This does not mean to imply that particular assessments about the appropriateness of actions cannot or should not be made. Yet, conflicting statements made about the abstracted acceptability of carpet bombing and cluster bombs are likely to leave a number of questions unaddressed about how specifically people make sense of such activities. These conventions of language need to be scrutinized as to how statements are formulated and advanced.

A third issue is that, in examining claims about weapons, the matter in which a technology and its context are discussed is worth close inspection. As has been mentioned above, it is often said that the effects or appropriateness of technology vary by context. After the events of 11 September, few would have failed to take notice of how seemingly mundane objects – shaving razors, screwdrivers, nail files – took on wholly different interpretations in relation to air travel. While the noting of context-dependent qualities of technology is necessary, just doing this alone is a limited approach. Following wider arguments in the social sciences about how meaning is negotiated in everyday life,[52] in this book I treat technologies and their contexts as mutually elaborating each other. This is to say that our understanding of technology and its context are made in relation to each other.

To explain, an account of technology or anything else relies on context as a way of understanding its meaning. Because descriptions are never exhaustive of the events they describe, statements must be made sense of in relation to certain assumptions. This requires considering a statement against a background of who said it, where and when, what was being accomplished by saying it. To speak of plastic bullets as a non-lethal option in policing Northern Ireland, say, in spite of the deaths due to its deployment there, can mean to suggest that many more deaths would have taken place had this option not been available, given the need of the security personnel to respond to public disorder and attacks made upon them. In this sense, descriptions of technology help elaborate a particular understanding of circumstances of their use.

Not just that the characteristics of a technology are given meaning by a particular setting, but also they give meaning to the setting. Particular accounts of technology make observable particular features of a setting. Take for instance the case of blinding laser weapons mentioned in the last chapter. In pressing for and achieving restrictions on the development of such devices, critics had to define the capabilities of the technology and the context of their use in relation to one. Determinations about the acceptability of blinding lasers rested on an interpretation of the context in which they should be understood (in general warfare), while at the same time, the important aspects of the context for evaluation (in this case the type of casualties incurred rather than the intent of military forces) derived from a sense of how the laser technology would operate in that context (it would be highly effective in causing blinding). Alter the technology or the effects attributed to blinding lasers, and features of the context deemed relevant are likely to change and vice versa. Thus, a description of technology evokes a context that it is made sense of and, in turn, the results of this process elaborate the specific meaning of the descriptions.

If understanding between parties can be achieved through the mutual definition of context and technology (as in the UN blinding laser negotiations), then a lack of it may also pose a significant impediment. Because accounts rely on some level of shared agreement regarding context and technology, it is possible for radically different meanings to be advanced. As will be explored more in Chapter 6, those campaigning against the use of plastic bullets in Northern Ireland do not define them as relatively non-lethal force options given the threats faced to security forces, but rather as an instrument of intimidation often used in situations where there is little risk of loss of life or property, which in turn leads to an altogether different assessment of the lethality of the technology. Just as the likely assessment one makes about the acceptability of 'cluster bombs' or 'carpet bombing' depends on certain assumptions one brings to bear, statements about non-lethals hinge upon assumptions about the context of their use and vice versa. In this sense, the claims made about the context of use (just like claims about technology) should be considered as an argumentative resource. Questions need to be asked about what context is implicitly or explicitly drawn upon in statements about the merits of non-lethal weapons and what this means for the way such technologies are portrayed.

We can note further, though, that accounts of technology and its context are in turn made sense of by audiences where some types of claim are considered more credible than others. The merits of particular accounts are defined in relation to specific settings and these are in turn defined by the accounts on offer. In the case of debates about blinding lasers, for instance, the type of argument made against their development rested on expert medical and scientific grounds, rather than simply on ethical lines of argument about the reprehensibility of certain weapons. Following Prokosch's general analysis and another first-hand account of the talks, it seems reasonable to assume that this was because, in the world of international arms discussions, 'emotional' arguments based on ethical reasoning 'alone' hold little sway.[53] Expert scientific and medical evidence was taken as credible within the setting of the United Nations arms-control forum, and that served to define the forum as a credible one for discussing what constituted acceptable means of warfare.

The claims made about the use of weapons depend on what is known about such events: what is known about the technology, what is known about the context, and what is known about who has such knowledge. With regard to the last aspect, the ability of individuals and organizations to comment on events is not uniform. When certain forms of expertise are considered more credible then others, then how that expertise is distributed within and across organizations is a crucial concern. In later chapters, considerations about the credibility of particular forms of argument and the distribution of credible expertise are given detailed consideration. Figure 2.1 provides a further illustration of the way in which technology and its context define each other and why this process should be seen in relation to the knowledge at hand.

FIGURE 2.1

AN ILLUSTRATION OF HOW LETHALITY, CONTEXT, TECHNOLOGY AND
KNOWLEDGE ARE MUTUALLY DEFINED AND WHY THIS MATTERS

As an organization that provides medical and humanitarian assistance in many conflict regions, the International Committee of the Red Cross has documented the casualties and fatalities incurred in warfare. As part of this, two of its medical officers (Robin Coupland and David Meddings) conducted a literature review of the number of people wounded and killed by conventional weapons and firearms in armed conflict as well as civilian mass shootings between 1929 and 1996.[54]

They found the mortality inflicted by firearms depended on the context in which they were used. In most situations of warfare, the wounded to killed ratio varied between 1.9 and 27.8. However, when those shot were in a confined space or when they were unable to defend themselves (such as in civilian mass shootings), the wounded to killed ratio was generally lower than 1. In other words, the percentage of people killed in armed conflict was much lower than that for civilian mass shootings.

Coupland and Medding argue that these results can serve as an indicator for when war crimes, such as the execution of wounded combatants or prisoners, take place. In a given situation when the number of wounded by conventional weapons equals the number killed, then this provides strong evidence of deaths caused by execution rather than by armed battle. They advise future attention to this ratio by the media, states and others.

In practice, though, it seems unlikely that such proportional figures *alone* would prove decisive in recognizing war crimes. The attempt is being made to define a context (one in which war crimes have taken place) on the basis of the effects of technology. The basic issue is that any ratio figures have to be interpreted in relation to a particular situation. This leads to at least two difficulties. First of all, as the authors themselves point out, the number of wounded in conflict is often difficult to determine as these individuals may not necessarily seek or receive medical treatment and such treatment might not be recorded. In contrast, assessing the number of dead is generally more straightforward. In many ways, the difficulty of counting the wounded is already accounted for by the wartime ratio of 1.9 and 27.8, which figures themselves already probably underestimate the number wounded in any given situation. However, in any particular instance where war crimes are alleged because of a low wounded to killed ratio (say below 1), pressing questions will need to be asked whether in this *particular* situation there are reasons why a proportionally low number of wounded to killed were counted. Perhaps this will be due to poor medical infrastructures.

Second, beyond limitations of the knowledge at hand, in any particular instance, further questions will need to be asked as to whether there were reasons for the low wounded to killed ratio. Did this, for instance, reflect the type of physical conditions in which combat was fought (say in an urban setting that featured close-quarter combat, thereby increasing the overall lethality) or the willingness of individuals to keep fighting while wounded rather than surrender? In other words, the effects of technology cannot solely define the context, but an understanding of the context must exist as well. This understanding in turn helps define the meaning of the effects. In debates about whether wartime atrocities have taken place, one of the likely key considerations would no doubt be the moral standing (or lack of it) of those accused. It seems reasonable to suggest that those forces already suspected of having committed, or deemed likely to commit, war crimes would be the ones against whom low wounded to killed ratios would initiate accusations of war crimes. It further seems realistic that such assessments of the behavior of particular forces and the way wounded to killed ratios are interpreted by the media and outsiders to a conflict would vary depending upon which parties to a conflict are supported politically and which are not.

The previous argument is not meant as refutation of the possibility or importance of thinking about war crimes in armed conflict through wounded to killed ratios. Rather, it attempts to outline something of how ratios are likely to come to bear in the determination of such crimes. Once this is acknowledged, we can understand how arguments are formed and the basis for disagreements.

The wounded to killed ratio also has relevance for the evaluation of non-lethal weapons. Coupland and Medding suggest that the deployment of non-lethals, such as entrapment devices and calmative agents, may in the future make it increasingly difficult for individuals to take cover or otherwise evade fire. Given that such devices are likely to be used in conjunction with more conventional weapons (and thus act as pre-lethal weapons), they may well lead to higher numbers of fatalities and this possibility should be scrutinized as part of international humanitarian law. Again, here the context defines the weapon and the weapon defines the context.

CONCLUSION

This chapter set out to address the basic issue of what constitutes a non-lethal weapon. Rather than this being a matter of mere words, contrasting definitions and characterizations point to alternative ways of making sense of the aims, effects and scope of this class of weaponry. The manner in which non-lethal weapons are differentiated from lethal ones for many proponents, for instance, rests on intention as much as, if not more than, actual effects. In other words, definitions suggest what counts as possible and plausible actions. Where the importance of these concerns about the meaning of weapons particularly comes into play is in evaluations of what types of control ought to be placed upon them.

Through considering such definitional issues, this chapter has also set out something of the orientation taken to statements about non-lethal weapons. The approach adopted here is one of analytical skepticism to the claims and counterclaims made. Even in the discussion of non-lethals that has taken place so far, it is clear that there is scope for contention about the capabilities and desirability of this technology. Skepticism entails giving the claims made some level of scrutiny rather than merely using them as a resource in advancing particular assessments. Furthermore, evaluations of technology are bound up with an understanding of the contexts in which they are utilized. As such, these contexts and the relation between technology and context need to be given attention. So, too, does the distribution of 'credible' expertise in making assessments. With these points in mind, let us now turn to a survey of the range of technologies for sale under the non-lethal banner.

NOTES

1. See e.g. Pilant, L., *Less-than-Lethal Weapons* (Washington, DC: International Association of Chiefs of Police, 1993).
2. Toffler, A. and Toffler, H., *War and Anti-War* (New York: Little, Brown and Company, 1994).

3. See e.g. Rothstein, L., 'The "Soft Kill" Solution', *The Bulletin of Atomic Scientists*, March/April (1994), 4–6; Federation of American Scientists, 'Non-lethal Weapons', *Journal of the Federation of American Scientists*, 48, 1 (1995); and Spinney, L., 'A Fate Worse than Death', *New Scientist*, 18 October (1997), 26–7.
4. See Egner, S., Shank, E., Wargovitch, M. and Tiedemann, A., *A Multi-Disciplinary Technique for the Evaluation of Less Lethal Weapons* (Aberdeen, MD: US Army Human Engineering Laboratory, 1973).
5. Anon, 'Nonlethal Weapons Give Peacekeepers Flexibility', *Avaiation Week and Space Technology*, 7 December (1992), 50.
6. See Wright, S., *An Appraisal of Technologies of Political Control*, Report to the Scientific and Technological Options Assessment of the European Parliament, PE 166.499 (Luxembourg: STOA, 1998).
7. Lewer, N. and Schofield, S., *Non-lethal Weapons* (London: Zed, 1997).
8. See e.g. Steering Group for Patten Report Recommendations 69 and 70 Relating to Public Order Equipment, *A Research Programme into Alternative Policing Approaches Towards the Management of Conflict*, December (Belfast: Northern Ireland Office, 2001).
9. Aftergood, S., 'The Soft-Kill Fallacy', *The Bulletin of the Atomic Scientists*, September/October 1994, 40–5.
10. Omega Foundation, *Crowd Control Technologies – Technical Annex*, Report to the Scientific and Technological Options Assessment of the European Parliament, PE 168.394 (Luxembourg: European Parliament, 2000).
11. Alexander, J., 'An Overview of the Future of Non-lethal Weapons', *Medicine, Conflict, and Survival*, 17, 3 (2001), 180–93.
12. See US/UK NLW Urban Operations Wargaming Program, *US/UK Non-Lethal Weapons/Urab Operations Executive Seminar – Assessment Report*, 30 November (London: Ministry of Defence, 2000), 3.
13. Lamb, C., *Non-lethal Weapons Policy:Department of Defense Directive*, 1 January 1995, p. 1.
14. Bunker, R. (ed.), *Nonlethal Weapons*, INSS Occasional Paper 15 (CO: USAF Academy); Griffioen-Young, H. and Janssen, H., 2001, 'Psychological Operations', in *Non-lethal Weapons*, proceedings of the 1st European Symposium on Non-lethal Weapons 25-26 September 2001, Ettlingen, Germany (Postfach: ICT, 2001); Council on Foreign Relations, *Non-Lethal Technologies: Military Options and Implications. Report of an Independent Task Force* (New York: Council on Foreign Relations, 1995).
15. Morris, C., Morris, J. and Baines, T., 'Weapons of Mass Protection', *Airpower Journal*, 9, 1 (1995).
16. New Scientist, *New Scientist*, 11 December (1993).
17. Amnesty International, *Imported Arms used in Israel and the Occupied Territories with Excessive Force Resulting in Unlawful Killings and Unwarranted Injuries* (London: Amnesty International, 2000).
18. For a vivid illustration of this see Brunner, J., *Stand on Zanzibar* (London: Millenium, 1986).
19. Alexander, J., *Future War* (New York: St Martin's Press, 1999), pp. 92–4.
20. Becker, J. and Heal, C., 'Less-than-Lethal Force', *Jane's International Defence Review*, February (1996), 62–4.
21. Taken from SIPRI, *Anti-personnel Weapons* (London: Taylor & Francis, 1978).
22. International Committee of the Red Cross, *The Medical Profession and the Effects of Weapons* (Geneva: International Committee of the Red Cross, 1996).
23. Committee on the Administration of Justice, *Plastic Bullets: A Briefing Paper* (Belfast: Committee on the Administration of Justice, 1998).
24. Taken from Doubet, M., *The Medical Implications of OC Spray* (Millstadt, IL: PPCT Management Systems, 1997).
25. Ijames, S., 'Less-Lethal Projectiles', *The Tactical Edge*, Fall (1996), 76–84.
26. Burrows, C., 'Operationalizing Non-lethality', *Medicine, Conflict, and Survival*, 17, 3 (2001), 260–71.
27. Granfield, J., Onnen, J. and Petty, C., *Pepper Spray and In-custody Deaths* (Alexandria, VA: International Association of Chiefs of Police, 1994).

28. Prokosch, E., *The Technology of Killing* (London: Zed, 1995).
29. E.g., Council on Foreign Relations Independent Task Force, *Nonlethal Technologies* (Washington, DC: Council on Foreign Relations, 1999).
30. Lovelace, D. and Metz, S., *Nonlethality and American Land Power* (Carlisle, PA: Strategic Studies Institute, U.S. Army War College, 1998).
31. Alexander, 'An Overview', 181.
32. Heal, Lt. Sid, 'The Evolution from "Non-Lethal" to "Less-Lethal"' [cited 20 March 2001]. Available from www.airtaser.com/Web_2000/Feb/SidHeal.htm
33. As Jürgen Altmann pointed out to the author, this is not necessarily a valid distinction. For instance, nets and tasers primarily target one's ability to act.
34. For some instances of this see Amnesty International, *Annual Report 1999* (London: Amnesty International [International Secretariat], 1999).
35. Bunker, *Nonlethal Weapons*.
36. Lovelace and Metz, *Nonlethality and American Land Power*.
37. Honigsbaum, M., 'Arms Firm Avoids Export Ban', *The Observer*, 17 January (1999), 1.
38. Pasternak, D., 'The Pentagon's Quest for Nonlethal Arms is Amazing. But is it Smart?', *US News and World Report* (1997); Becker and Heal, 'Less-than-Lethal Force'; and Alexander, 'An Overview'.
39. Anon, 'Nonlethal Weapons Give Peacekeepers Flexibility', 50–1.
40. Spinney, L., 'A Fate Worse than Death?', *New Scientist* (18 October, 1997), 26–7.
41. See for instance Grint, K. and Woolgar, S., 'Computers, Guns and Roses', *Science, Technology, and Human Values*, 17, 3 (1992), 366–80; Woolgar, S. and Cooper, G., 'Do Artefacts Have Ambivalence?', *Social Studies of Science*, 29, 2 (1999), 433–45; and especially Grint, K. and Woolgar, S., *The Machine at Work* (Cambridge: Polity, 1997).
42. For an earlier version of some of the arguments of this book that elaborate the implications of this approach for theories of technology, see Rappert, B., 'The Distribution and the Resolution of the Ambiguities of Technology, or Why Bobby Can't Spray', *Social Studies of Science*, 31, 4 (2001), 557–91.
43. By this I mean either the actual using of or threatening to apply force. For some (arguably contestable) figures see Hepburn, J., Griffin, M. and Petrocelli, M., *Safety and Control in a County Jail* (Tempe, AZ: Arizona State University, 1997).
44. Woolgar, S. 'The Turn to Technology in Social Studies of Science', *Science, Technology, and Human Values*, 16, 1 (1991), 20–50.
45. Northam, G., *Shooting in the Dark* (London: Faber & Faber, 1988).
46. Woolgar and Cooper, 'Do Artefacts Have Ambivalence?', 443.
47. Grint and Woolgar, *The Machine at Work*, 368.
48. Woolgar, S., 'Virtual Technologies and Social Theory', in R. Rogers (ed.), *Preferred Placement* (Maastricht: JanVan Akademie Editions, 2000), p. 175.
49. See Latour, B., *Science in Action* (Milton Keynes: Open University Press, 1987).
50. See Grint and Woolgar, *The Machine at Work*, p. 100 and Coulter, J., *Mind In Action* (London: Polity, 1989).
51. As for instance in 'public debating' forum such as Question Time, 2001, *Question Time- BBC Productions*, aired 25 October.
52. Here I refer to the work of ethnomethodologists in elaborating the concepts of reflexivity and indexicality. For an introduction to such work see Heritage, J., *Garfinkel and Ethnomethodology* (Cambridge: Polity, 1994). These notions have been applied to technology in, for instance, Grint and Woolgar, *The Machine at Work*.
53. Prokosch, *Technology of Killing* and Doswald-Beck, L., *Blinding Laser Weapons*, Papers in the Theory and Practice of Human Rights, Number 14 (Colchester: University of Essex, 1995).
54. Coupland, R. and Meddings, D., 'Mortality Associated with Use of Weapons in Armed Conflict, Wartimes Atrocities, and Civilian Mass Shootings', *British Medical Journal*, 319 (1999), 407–10.

3

Tools of the Trade

Since the early 1990s, support for non-lethal weapons has grown apace. While aspirations have generally outstripped working technologies, today, across much of the Western world and beyond, steps are being taken to fund weapons at all stages of development. The growth in interest has entailed the creation of both organizations to sponsor research and managing bodies for these organizations. The US Joint Non-Lethal Weapons Directorate, for instance, co-ordinates joint activities across the armed forces. As a testament to the rising importance of this weaponry and what, on the face of it, can only be taken as a sign of the secure footing of such programs, the Directorate website has a page exclusively devoted to relevant non-lethal weapon (NLW) acronyms.[1] The hundreds of acronyms listed include various terms of military jargon, institutional reference points and weapon abbreviations.

The existence of such a site also suggests the diversity of technical options being considered. Although 'non-lethal' serves as a rallying term for those who support and oppose certain initiatives, the term in itself does little justice to the array of technologies on offer. This chapter provides an overview of current efforts, what devices are advertised and who is supporting them. In compiling an overview of major developments, I draw on and in some places update a number of existing surveys of non-lethal weaponry.[2] Following the points made in previous chapters, an indication of the operational contexts of use, as well as the possibilities and pitfalls associated with particular options, are mentioned. The reader will no doubt think of situations in which particular non-lethal weapons might be deployed and the consequences therein. The second purpose of this chapter is to establish the main weapons under consideration in the book.

'FIRST GENERATION' WEAPONS

[I]t is a general principle that the police, and (if they are called upon) the military, should employ only the minimum degree of force necessary to restore order or protect life and property in the event of riots, and that recourses should be had to the use of firearms only as a last resort. In the view of the serious consequences which result from firing upon civilians, it is I feel important that alternative methods for the dispersal of crowds should be continuously studied. It may, for example, be possible to evolve tactics or techniques which, though not necessarily of general application, may be effective in particular circumstances or may be specially suited to local conditions in individual Colonial territories ... One of the most effective and humane weapons available against rioting crowds is tear smoke.[3]

So writes Arthur Greech Jones, the British Secretary of State for the Colonies, in a then classified circular to colonial governments during 1948. In instances where the recourse to force is deemed necessary, 'tear smoke' (what is now generally referred to as tear gas) can forestall the use of firearms. The quote indicates the long-standing search even in modern security forces for what are referred to today as non-lethal weapons. This section describes some of the technologies that were pursued in the twentieth century, prior to the reinvigorated interest in this class of weapons that began around 1990. This is done by considering the prominent options and programs of research in the United Kingdom and the United States, the sites of many relevant innovations.

Besides indicating the history of interest in non-lethals, the quote from Jones suggests what is at stake in the stories told about this topic. That tear smoke, let alone firearms, would be seen as necessary suggests that events are volatile and presumably violent. The characterization of tear smoke as a humane alternative was done in relation to firearms. As opposed to the logical need for finding new means of force that might be shared by the intended audience of this circular, for others the reference to restoring order in relation to rioting in the colonies would be likely to bring a different interpretation about the likely legitimacy of employing tear smoke.

The use of so-called tear gases (labeled as such because one of their effects is the rapid production of tears) as non-lethal forms of weaponry dates back to the First World War. Tear gas was used to force opponents to don gas masks, thus reducing their effectiveness and inducing psychological stress. With the economic and social turmoil that followed the First World War, police forces also became interested in such weapons. A tear gas canister fired into a crowd was meant to force individuals to flee because of the pain of remaining. One of the widely agreed cautionary points about tear gases is that they build up in enclosed spaces in a manner that makes them less than harmless.

Much of the attention after the First World War in the United States and Britain centered on CN (chloroacetophenone). CN produces a burning, irritating and stinging sensation in the throat, nose and eyes thereby leading to profuse crying and salivation. It was actively promoted as a means of quelling rioters without resorting to firearms. In 1936, the British government approved CN for use in colonies such as Kenya, Jamaica and Trinidad. By the late 1940s, however, doubts were expressed about its effectiveness. Experience with the chemical enabled some to build up a resistance to it, and its judged low potency meant it was fairly easy to counter (for instance, by shielding one's face).

According to Evans, this and the perception of growing hostilities around the world and at home led British scientists at defence establishments such as Porton Down to refine a new incapacitating agent, CS (o-chlorobenzylidene malononitrile).[4] For a given dosage it is considered much more potent than CN and thus able to harass individuals at lower concentrations. In 1956, CS smoke was deployed in Cyprus and British Guyana. Its use became fairly widespread across the colonies as Britain attempted to remain and then pull out of its positions. During the 1950s, the United States and Britain shared technical information about the agent. By the 1960s, the US Army became active in finding weapons for use in counterinsurgency operations that would not maim or kill. CS gas figured prominently in the Vietnam War (see Chapter 6).

A third major chemical incapacitant is CR (dibenz [b.f.]-1-4-oxazepine), first synthesized in 1962. Like other irritants, it causes severe eye and skin pain, but at still lower concentrations than with CS. The health concerns associated with CR have been of particular concern to many governments and arguably for this reason its use has not been as widespread as CN and CS. Such concerns did not stop its internal use in South Africa by defence forces during the apartheid rule.[5]

Irritants have not been the only chemical means sought for incapacitating or disabling individuals without death. Both Britain and the United States in the 1950s and 1960s had active programs to test the properties of various chemicals intended to target the mental state of opponents.[6] In the 1950s, the US Chemical Corps experimented with the disabling qualities of LSD, mescaline and marijuana. On the basis of initial American efforts, Britain began LSD research in the 1950s.[7] Ultimately, LSD was deemed unworkable because of problems with dispersing it, the large quantities required for ensuring its effectiveness in battlefield conditions, its illegal status, and its highly variable effects.[8] The US Chemical Corps continued its research about disabling weapons, including the psychotropic drug BZ (3-quinuclidinyl benzilate). In the 1960s, now rather notorious experiments on the effects of LSD were conducted in the United States by giving it to university students and prisoners without their knowledge. What information existed in the public at the time explicitly billed psycho-chemicals as means of incapacitating without killing; the suggestion was even made that they were a more humane alternative than the atom bomb.[9]

Kinetic-energy non-lethal weapons work by imparting blunt trauma force, ideally causing minimum long-term injury. British authorities in Hong Kong deployed wooden 'bullets' launched from a specially designed gun in 1958. In the 1970s, first rubber and then plastic bullets were introduced in Northern Ireland. In many respects, devices such as vehicle-mounted water cannons function as kinetic-energy weapons by imparting a force to those targeted. Injuries are also produced by knocking individuals to the ground or knocking loose objects into them. Such cannons have figured in the policing of protests and riots in Europe for many years.[10]

A wide range of technologies have been developed that use electrical currents to incapacitate or shock individuals by interfering with nerve signals within the body: hand-held stun 'guns', stun batons and stun shields have electrodes that administer energy at the flip of a switch. These are battery-powered and give out a low-current, high-voltage impulse that is supposed to incapacitate limbs or whole bodies. Some such devices contain additional features such as enhanced sparks or small loudspeakers that enable greater visual and audio effect. These particular forms of stun weapons are predated by related technologies, such as cattle prods, that were used on humans. A variety of equipment allows for administering shocks over a distance. One prominent sub-class of such equipment is TASER® technology, first introduced in 1970. The name derived from the children's story 'Thomas A. Swift's Electrical Rifle'. TASERs use propelled probes connected by wires to a power source.

Previous non-lethal weapons predominantly fell into categories of chemical, kinetic and electrical devices. There were, however, other types of device. A so-called 'squawk box' was supposed to have been developed for crowd control in Northern Ireland. It reportedly consisted of two loudspeakers with an output of a few hundred watts that each produced slightly different frequencies just at the hearing threshold level of humans. This box was said to induce feelings of nausea and panic, though there seems to have been some confusion or deliberate obfuscation regarding what it did and did not do.[11] In any case, this acoustic device never became part of the standard repertoire of any force, though it is still referred to today as a viable option.[12] So called 'flash-bang' or 'distraction' grenades (devices that produce an intense flash of light and single or multiple loud bangs) have been used by police and military forces for a number of years.

In 1974, the Security Planning Corporation produced one of the earliest publicly released assessments of non-lethal weapons for the US National Science Foundation.[13] American interest in the application of such technologies in policing increased significantly following the protests and civil disturbances of the late 1960s. While noting potential problems with the use of chemical, electroshock, kinetic and other technologies in police enforcement, the report advocated their continued development. The Security Planning Corporation advised further research on the suitability of different contexts of use as well as the physiological, psychological and social effects of such weapons. The report also noted the importance of a range of considerations about training, tactics and community relations in decisions about the appropriate responses. Defense-related laboratories in the early 1970s attempted to do just this by elaborating a systematic set of procedures to evaluate the desirable and undesirable effects of lethal and non-lethal weapons, covering a comparative assessment of both the medical and physiological consequences of each weapon type, together with an evaluation of their public acceptability.[14]

RE-EMERGING ATTENTION TO NON-LETHAL WEAPONS

Although the gradual development and deployment of non-lethal weapons continued throughout the 1970s and 1980s, there was little in the way of comprehensive support programs to foster such technology in much of the Western world. This began to change around 1990. Lewer and Schofield provide an overview of the emergence of non-lethal weapons funding in the United States during the early 1990s.[15] Herein, the growing attention to this weaponry is seen as a response to many post-Cold War dilemmas facing policy-makers, including the establishment of a rationale for maintaining expenditures on military research and development (R&D). In the post-Cold War environment, institutes such as the Department of Energy national laboratories, with their concentrations on nuclear weapons and other advanced technologies, were threatened with extinction due to the declining relevance of such skills. With a shifting emphasis in US research policy toward meeting strategic civilian priorities, new justifications had to be offered for weapon expenditures. The interest in non-lethals emerged out of the focus in the Clinton administration on so-called 'dual use' R&D that would have civilian and defense applications. Whether one interprets this new rationale as a sensible shift in resource allocations or a perpetuation of career and funding paths in need of reform, by the early 1990s, much of the growing interest in non-lethal weapons formed from cutting-edge defense establishment organizations.

39

The rethink conducted regarding US national R&D strategies paralleled, and to some extent drew from, re-evaluations of security policy. As opposed to Cold War strategies that pitched East versus West in a bipolar power struggle, commentators in the early 1990s pointed to new, and potentially more destabilizing, security challenges. Operations in Bosnia and Somalia revealed the complexity of military interventions involving extended contact with a divided population. Emphasis was placed on separatist and counter-insurgency wars, border disputes, ethnic and religious violence, *coups d'état*, national security and counter revolutionary operations. Advocates of non-lethals, such as Janet and Christopher Morris and John Alexander (the former director of the Los Alamos Disabling Technologies Program), played major roles in giving a greater spotlight to non-lethals. These technologies were billed as providing much-needed flexibility in the use of force for peacekeeping missions and resolving regional conflicts. Such demands in the international arena were said to intertwine with public order and routine domestic policing concerns.

Although aborted attempts were made under the Bush presidency to provide non-lethal weapons with an institutional footing, it was not until 1993–94, under the Clinton administration, that a high-level steering group was set up through the Under Secretary of Defense for Acquisition and Technology and the Office of the Assistant Secretary of Defense for Special Operations and Low Intensity Conflict. Individual US armed services (Army, Navy, Air Force, Marines) conducted significant research into different options, though due to secrecy concerns, the full nature and level of activities is not publicly known. In 1994, the Department of Defense signed a memorandum of understanding with the Department of Justice to collaborate on non-lethal options that could serve police and military forces.

In 1997, the Joint Non-Lethal Weapons (JNLW) Program was established to co-ordinate actions between the services as expressed in a memorandum of agreement signed by the armed forces. Through its 'Master Plan', the program's Directorate seeks to define the core capability requirements and guiding principles underlying non-lethals funding in the United States. The capabilities sought to include controlling crowds, denying vehicles and persons access to areas, clearing spaces of vehicles and persons, incapacitating individuals and disabling equipment. These might feature in military operations such as evacuations, peacekeeping, anti-terrorism, as well as general warfare. To these ends, the Directorate funds concept explorations, the design of prototypes, and the acquisition of specific technologies.

Much to the disappointment of some, the JNLW Program only facilitates joint programs.[16] Individual services still conduct their own initiatives and much less is publicly known about these activities. This situation makes it difficult to judge the scale of research and development efforts under way. The JNLW Program had a budget of $30 million a year prior to 11 September 2001. Although discussions are under way to increase this significantly post-11 September, this scale of funding is relatively small by military standards. Yet, presumably, this is only a fraction of the total funding available for non-lethals research in the US military. Furthermore, as is detailed below, many major commercial defense contractors are now beginning to establish non-lethal development programs.[17]

In terms of American law-enforcement agencies, the National Institute of Justice seeks to identify, develop and evaluate what it calls less-than-lethal devices. As with military requirements, police seek means of restraining and incapacitating individuals as well as controlling crowds. The potential of this technology became the focus of police and popular attention after the beating of Rodney King in Los Angeles, and deaths caused to the Branch Davidians outside of Waco, Texas, in the early 1990s. Table 3.1 lists the major grants funded by the Institute's Office of Science and Technology.[18] Several of these projects will be considered in this book.

In addition, various structures have been set up to assess the medical, psychological and social implications of technologies. The JNLW Directorate has established the quasi-independent Human Effect Advisory Panel to provide evaluations of the range of considerations for specific non-lethal technologies, such as acceptable dosages of chemical and kinetic-impact levels. The Non-Lethal Technology Innovation Center at the University of New Hampshire identifies and develops materials and other technologies for use in non-lethal weapons. Priorities of the Center in 2001 included research into malodors, acoustic weapons, high-power electromagnetic devices, chemicals for degrading rubber, and fuel contamination substances.

As part of efforts to promote the development of non-lethal weapons, the US government has signed memorandums of understanding with Britain and Israel. These bilateral agreements entail the exchanging of classified and non-classified information on strategies and equipment. Britain and the United States also conduct various wargames with non-lethal weapons, including for scenarios such as humanitarian assistance, peacekeeping and urban combat.[19] A program of joint conferences between Russian and American officials on the application of non-lethals in anti-terrorist and peacekeeping operations was established in the late 1990s.

While much of the recent impetus behind non-lethals stems from the United States, programs of support are now in place in many countries. In October 1999, the North Atlantic Treaty Organization adopted a

TABLE 3.1
LESS-THAN-LETHAL GRANTS FUNDED BY THE US NATIONAL
INSTITUTE OF JUSTICE

Title	Start Year
Public Acceptance of Police Technologies	1993
Aqueous Foam System	1994
Evaluation of Oleoresin Capsicum and Stun Device Effectiveness	1994
Less-Than-Lethal Technology Assessment and Grant Transfer	1995
Law Enforcement Technology, Technology Transfer, Less-Than-Lethal Weapons Technology and Policy Liability Assessment	1996
Ring Airfoil Projectile System for Less-Than-Lethal Application	1997
Health Hazard Assessment for Kinetic Energy Impact Weapons	1997
Evaluation of Oleoresin Capsicum	1997
Pepper Spray Projectile/Dispenser	1997
Laser Dazzler Assessment	1998
Impact of OC Spray on Respiratory Function in the Sitting Prone Maximal Restraint Positions	1998
Evaluation of Vehicle Stopping Electromagnetic Prototype Devices	1998
Biomechanical Assessment of Non-Lethal Weapons	1998
Research and Establish a Computerized Database of Firearm Delivered Less Lethal Impact Munitions	1998
Evaluation of the Human Effects of the Sticky Shocker Topic	1998
Preliminary Characterization and Safety Evaluation of Defence Technology's OC Powder	1999
Electromagnetic Prototype Device – Phase III	1999
ROAD SENTRY Vehicle-Stopping Prototype Electrostatic Device	1999
Applicability of Non-Lethal Weapons Technology in School	1999
Less-Than-Lethal Ballistic Weapon	2000
Less-Than-Lethal Equipment Review	2001
Pepper Spray's Effects on a Suspect's Ability to Breathe	2001

policy on non-lethal weapons.[20] In Britain, the Defence Evaluation and Research Agency (recently partly privatized into QinetiQ), as well as the Police Scientific Development Branch, give support for non-lethal research. The Netherlands has set up a research program conducted by TNO Defence Research Organization. Funding under this program includes next-generation technologies such as low frequency acoustic-energy weapons designed to incapacitate individuals, psychological operations, high-powered microwaves for damaging communications equipment, lasers that dazzle, kinetic-energy weapons and capture nets. A recently established Europe-wide symposium on non-lethal weapons enables the exchange of expertise.

In sum, non-lethal weapons research efforts that in the past were fragmented and limited in scope are being given greater resources and are becoming co-ordinated. With public financial support for research, an expanding manufacturing base and various national and international conferences, the networks associated with non-lethal weapons seem set for expansion. Let us consider, then, the new technologies envisioned.

THE NEXT GENERATION AND BEYOND

Included within the range of current and planned non-lethal weapons are refinements of past weapons and alternative means of dissemination, as well as altogether novel options.

Kinetic

Varieties of kinetic-energy weapons, or what are sometimes called low-impact blunt munitions, form a key type of non-lethal weapon in law-enforcement and military peacekeeping settings. They are used to disperse, control and incapacitate individuals in incidents of barricaded armed suspects, violent encounters and civil disturbance. In policing situations, kinetic weapons also aid in subduing suicidal or mentally ill people armed with knives or other weapons.

A wide range of governments and firms fund the development of non-lethal munitions as well as their launchers. Kinetic weapons differ in their energy levels, whether for direct or indirect targeting and for single or multiple shots. Each variant is supposed to fulfill some particular operational or safety requirement. So, bean bag canvas rounds are meant to distribute impact energy across the whole contact area. The potential for such bags to ricochet in unpredictable ways if they hit a person edge-on from the side has led to the development of a 'sock round' that contains a metal shot in an elongated pouch. A closely related type of projectile is the so-called sponge round, which consists of a hard rubber sponge that deforms on impact. Chemical irritants that are released during the deformation can be added to such munitions as well.

In the late 1990s, the US National Institute of Justice supported the development of the ring airfoil projectile; a ring of rubber that weighs about 30 grams with a 2.5-inch diameter. The projectile can be launched from an M-16 rifle and is meant to have a fairly level trajectory. The ring shape can be filled with chemical irritants or other materials. Developers trace interest in this technology to such diverse origins as the need to find a means of firing (lethal) fragmentation explosives along a flat trajectory to avoid jungle canopies in Vietnam,[21] as well as the shooting of students in war protests at Kent State University in 1970.

While most kinetic projectiles are designed to be fired from a 12-gauge shotgun or a specialized launcher (typically 37 and 40 mm), these are not the only means available. Hand-thrown grenades exist that release impact projectiles. In 2001, the JNLW Program funded the development of the Modular Crowd Control Munition, a variant of the claymore mine, that discharges rubber balls.

43

Continuing modifications have been made to water cannons. While earlier versions had considerable operational limitations associated with their maneuverability, storage potential and flow rate, over time their capacities have gradually expanded. As well, attempts are being made to devise portable liquid-projectile weapons to deliver kinetic-energy impacts. Portable water cannons come in shoulder and small wheelable containers. Manufacturers of the 'Liquid-Projectile Weapon' say it can be used to keep people at a distance without inflicting serious harm, thus reducing the 'CCN [Cable News Network] factor', or adverse publicity associated with killing or seriously injuring noncombatants.[22] High-voltage, low-current electroshock pulses can also be delivered via water cannon through the incorporation of a conductive fluid in the water stream.

Mechanical

Mechanical forms of non-lethal weapons include devices such as nets and barriers aimed at physically controlling or restraining movement. Primex Technologies, for instance, has devised a portable net system to stop vehicles.[23] Much of the effort in the area of mechanical non-lethals is given to finding means of launching nets and other restraints at a distance from those targeted. A potential problem with such stand-off devices is that the impact of fired nets and related flight stabilizing weights can cause serious injury, for instance, by contorting the neck. Standard nets can be laced with chemicals (for instance, sticky materials or irritants) or electrified to provide additional effects. Simple caltrops (spikes that can be spread along the ground) are also being devised for administering chemicals.

New materials for physical restraint are also being considered. In a rather comical example that conjures up images of the superhero Spiderman, the US Army Soldier Center is seeking to harness particular types of spider fibers for use in nets. In order to obtain sufficient quantities of such materials, molecular biologists are at work transferring spider genes into bacteria, mice and goats in order for them to produce spider silk in their milk. It is then the intention to spin this silk so it can be incorporated into personal netting entanglement mechanisms.[24]

Chemical

Chemical irritants, the so-called tear gases, continue to be used in the control and breakup of crowds. A variety of dispersal mechanisms, such as sprays, cartridges, grenades, shoulder slug canisters, mines and mortars, have been engineered recently.[25] While harassing chemical agents were deployed throughout the twentieth century in the form of gases or smoke, it is only in the last thirty years of the century that personal chemical incapacitant sprays were readily available in some countries, and only in the last few years that their adoption became widespread. Oleoresin capsicum (OC) is the prominent chemical agent in the United States in this regard. Unlike chemicals such as CS, it is not properly described as an irritant but is an inflammatory agent. It is associated with inflammation of the skin and nose, an intense burning sensation, shortness of breath, coughing, closure of eyes, pain, and psychological conditions such as panic.

The general type of the work conducted by the US Chemical Corps in the 1950s and 1960s continues, albeit under the effort to find advanced forms of riot-control agents. Although much of this work remains secret, psychoactive chemicals are being researched to target receptor sites in the brain that have been identified with causing anxiety, submissiveness and hallucinations.[26] Calmative agents are being investigated to induce sleep and other effects in individuals.[27] As with the psychotropic drugs of the earlier generation, a major operational problem associated with such disabling chemicals is how they can be dispersed. Currently, efforts are underway in the United States to turn microencapsulation technologies developed in the pharmaceutical industry and elsewhere into delivery devices for non-lethal chemicals. The chemicals can be released by mechanical crushing or through chemical reactions (say, by water from a water cannon).[28] As part of its then secret chemical and biological weapons program, South African Defence Forces sought to develop methaqualone (mandrax) and Ecstasy as crowd-control chemicals.[29]

A class of chemicals considered by some to hold much promise for the future is malodorants, or 'olfactory stimulation agents'. The use of 'stink bombs' is hardly new in warfare. In the Second World War, for instance, the US Office of Strategic Services developed foul-smelling chemicals. During the Vietnam War, the US Defense Advanced Research Projects Agency attempted but failed to find offensive smells that were culturally specific to the Vietnamese.[30] Today, efforts are being made to link particular smells to emotional responses such as pain and fear[31] that 'can be applied against any population set around the world to influence their behavior'.[32] In this sense, malodorants are not just supposed to be repugnant but to induce intense reactions. Again, the stated rationale is to find a means of dispersing crowds that poses minimal risk of fatalities.[33]

Various foam-like substances are under development as area-denial instruments and means of immobilizing people. Foam dispensers consist of a compact liquid and an expansion gas that are fitted into a container that is slung around the shoulder or otherwise made portable. Sticky foam consists of a toffee-like glue that adheres to surfaces contacted. Initially, it was hoped that this foam would act as a means of restraining individuals. For a number of years, the image of a human-like object covered in sticky foam served as a prominent icon in discussions of non-lethal weapons.[34] Its development was reported as imminent, several times.[35] The difficulty of removing the foam and the possible deaths and injuries associated with its use should the material come into contact with the eyes, mouth or nose, though, led to its anti-personnel applications being sidelined. The interest in sticky foam has now shifted to finding foams for denying easy access to buildings by sealing doors, windows and other entry points. Aqueous foam consists of bubbles that look like soapsuds. A type of aqueous foam was originally developed in Britain to fight coal-mine fires. Being covered in the foam reduces hearing and restricts movement. It was refined by US Sandia National Laboratories to function as a safety guard for military and nuclear installations.[36] By the 1990s, as part of the growing interest in non-lethals, attention was given to its potential in correctional facilities. Additives such as dyes or chemical irritants can be added to both the sticky and aqueous foams.

As part of efforts to deny access to areas and slow down the movement of persons and vehicles, 'slippery' anti-traction substances are being examined. Variations of household substances such as lard, oil and syrup, as well as more specialized chemicals, are being considered. The US Marines are currently taking the lead in coming up with a workable substance for a variety of environments, though the identification of a suitable chemical has been hard going.[37]

The US Army Edgewood Chemical Biological Center is exploring the possibility of disrupting the functioning of engines and other machinery through the addition of fuel additives or lubricant degradants.[38] This would enable equipment to be rendered obsolete without risk to human life caused through bombing. It has long been known, for instance, that adding water to gasoline can prevent vehicles from working. The range of chemical non-lethal weapons currently being envisioned also includes so-called supercorrosives, such as potent acids, which could render machinery unusable by dissolving metals and other materials.

Electroshock

Electroshock capabilities are finding their way into an increasing number of applications. Research supported by the US Defense Advanced Research Projects Agency and the National Institute of Justice has led to the production of the Sticky Shocker®, a wireless, self-contained projectile that has an electrical power source and short barbs (or adhesives) that make the projectile cling or 'stick' to individuals.

An American company called Tasertron is devising a non-lethal version of the anti-personnel mine. Once activated by sensors in the device or motion detectors, the taser mine sends out darts connected by wires to a power source within the mine. Individuals snagged by the darts receive repeated periodic shocks. According to the inventor, 'the subjects remain conscious and coherent but cannot control their limbs until the power has been turned off and they recover'.[39] A variety of operational settings for this type of mine are envisioned, such as border control, school protection and area control.

Biological

Employing biological microbes as a non-lethal weapon has attracted some attention.[40] Micro-organisms like bacteria could degrade or destroy equipment, damage crops, and eat away at explosives or key components of equipment such as sealings or circuits. The possibilities introduced by developing organisms that devour particular materials are said to be abundant and already possible, given the current biological knowledge and skills available in almost any country.[41] The coating of aircraft could be reduced in order to make them more easily detectable by radar and the integrity of roads could be degraded. A micro-organism that degraded oil could be placed in central refineries and thus spread quite easily. Alexander suggests that scientific advances mean that freeze-dried organisms could be introduced to a target (such as a fuel supply) along with means of revitalizing them.[42] Long after the initial introduction, the organisms could be reactivated with little possibility that those affected could trace the actions back to their origins. Toxins, peptides and cell-signaling molecules could be adapted through genetic engineering or otherwise to alter body temperature, mood and hormone release, and through this to function as incapacitating biological agents.[43]

Acoustic

As indicated in the example of the British squawk box above, the idea of acoustic devices serving as disabling weapons has been around for some time. US forces played loud rock music to annoy and dispirit Manuel Noriega in Panama in 1990 during his retreat in the Vatican Embassy. The irritating potential of listening to unpleasant sounds are well known outside situations of conflict. Current efforts to develop acoustic weapons seek other, but temporary (and hence non-lethal), psychological and physiological effects such as fear, disorientation, uneasiness and pain by targeting the ears and inner organs.

Significant interest has focused on infrasonic frequencies; those just below the level of hearing.[44] Infrasound is said to resonate with organs and thus interfere with the functioning of body organisms causing nausea, physical discomfort, disorientations, bowel spasms and even death, if used inappropriately. The potential exists for creating high-intensity, low-frequency acoustic beams with 'tunable' effects (that is, their intensity can be adjusted depending on the outcome sought) that pass through solid objects such as building structures.

A number of specific acoustic technologies have been discussed. For instance, in the past, Scientific Application and Research Associates of California reportedly worked on a system that compresses air and turns it into low-frequency infrasonic waves that act as acoustic 'bullets' for knocking people down without causing lasting physical damage.[45] US Primex Physics International Company has discussed a prototype acoustic blaster that utilizes a combustion detonation generator mounted on a vehicle to provide directional sound for area denial and crowd control.[46] The German Fraunhofer Institut Chemische Technologie is developing a mobile combustion-driven 'infrapulse-generator'.[47] The effects are said to be three-fold: the production of disruptive noise; the generation of low-frequency shock waves that resonate with body organs; and the creation of a unidirectional air ring of mechanical-impulse vortex that shoots out from the tubes. At present, this ring is said to be strong enough to rip glasses off individuals at a distance of tens of meters. Efforts to increase the intensity of the effects of this are being made, as are attempts to introduce chemicals such as irritants within the propagation of the ring.

Electromagnetic-Pulse Weapons

The possibility of disturbing modern electronic circuitry through a short-duration pulse of electromagnetic energy has created a significant amount of concern in recent years. Electromagnetic pulses (EMP) can be generated by converting the energy from conventional explosives or nuclear reactions into radio-frequency pulses. Such pulses can disturb or damage communications and information equipment by entering directly through antennas or indirectly through physical holes. The US military is said to have invested hundreds of millions of dollars into researching such technology.[48] EMP-type vehicle-immobilization systems are under development to stop vehicles traveling at speed.

A key concern with this technology centers on the vulnerability of modern civilian equipment in highly computer-dependent Western societies. Efforts are under way to assess the likely impact of different wavelengths and power levels. The potential costs for transportation and key infrastructures such as hospitals could be substantial, as could the ensuing social disarray.[49] While sophisticated, high-powered EMP weapons carried in bombs and other such devices might pose the greatest scale of danger, researchers from the Dutch TNO Physics and Electronics Laboratory have reported that, with electrical equipment available in any commercial store, it is possible to craft a vehicle-portable system that disturbs unprotected computer equipment, thereby causing a loss of data and no access to hard disk, as well as shutting computers down.[50] The unlikelihood of detecting or tracing such weapons opens up numerous possibilities for those wishing to wreak havoc.

Microwave

While the potential of non-lethal microwaves against humans has been speculated upon in the past, little in the way of material technologies have been put forward. This situation changed in 2001 when the US Marine Corps and US Air Force announced the operation of the Vehicle Mounted Active Denial System (VMADS), dubbed the 'people zapper'.[51] Hailed as 'the biggest breakthrough in weapons technology since the atomic bomb',[52] VMADS is reported to operate at 95 gigahertz and produce short invisible microwaves that vibrate water molecules in skin. The pain produced has been equated with that of touching a light bulb. Skin temperatures rise to 130 °Fahrenheit (54 °Centigrade) in two seconds. Like other electromagnetic weapons, the microwave beam moves at the speed of light, so travel is nearly instantaneous. Various employments for crowd control are foreseen: forces could 'engage a crowd from afar, directing two-second bursts of energy without risk of being overcome by the mob. When the beam is waved over the group, individuals would immediately experience intense pain, causing confusion and driving the crowd to disperse.'[53] Because the microwaves do not penetrate the skin very deeply, there are supposed to be few long-term implications, although the basis for such claims has not been made public at the time of writing. As the name implies, the weapon is supposed to be mounted on a vehicle, though this possibility is said to be some way off because of the current size limitations.[54] Research is also being conducted by the US Air Force into the use of high-powered microwaves against aircraft and satellites.

Rosoboronexport, Russia's state arms trading company, is reported to be developing non-lethal, microwave-beam weapons ('Ranets-E' and 'Rosa-E') for disrupting the electronics of weapons and radar equipment. The Ranets-E device is reported to disable precision weapons within a range of 10km through short pulses of a centimeter-wave-band microwave radiation. Rosa-E is said to jam radar systems. In an attempt to find customers as well as funders, Rosoboronexport is inviting interested parties to specify their requirements and fund necessary research and development expenses. Buyers of the devices are thus purchasing not just pieces of hardware, but the expertise of Russian scientists and engineers.[55]

Lasers

Lasers are currently incorporated into a variety of military equipment such as range finders and guidance systems. The potential of lasers for detecting and disrupting optical and electro-optical sensors, and thus disrupting weapon control systems, reconnaissance and communication systems, has generated considerable interest. The effects of lasers depend on characteristics of lasers, including their power, wavelength, duration, mode of radiation and distance.

Instant in travel, potentially invisible and able to produce graduated effects on targets, lasers are highly versatile. Much of the current 'non-lethal' interest in lasers relates to their potential to cause discriminate effects. The aerospace giant, Boeing, is developing what it calls the 'Advanced Tactical Laser'.[56] This chemical-oxygen iodine laser is said to be able to accurately place a four-inch spot at twenty kilometers when deployed in an aircraft or helicopter. Another laser option is a Pulsed Energy Projectile, reportedly referred to as the 'phasers on stun' technology by the director of the JNLW Program.[57] It works by explosively ablating surfaces of targets, thereby generating recoil that exerts a mechanical impulse. The tunable effects range from lethal to non-lethal; the latter are said to include causing 'shrapnel-less flash-bang, cutaneous peripheral afferent nerves (pain, susceptibility to chemical agents, lesions), cutaneous peripheral efferent nerves (temporary paralysis, choking, fibrillation), central nervous system (disorientation)'.[58]

As indicated in Chapter 1, non-blinding and low-energy-level lasers are being developed for other purposes. In Somalia, it was reported that a prototype rifle-mounted 'Saber 203 was used successfully to "tag" intruders in several engagements resulting in Somalian adversaries surrendering or retreating in every instance. While tactically successful, Saber 203 was not used directly to accomplish visual jamming, since it was not approved medically or legally for direct injection of glare.'[59] Lasers designed specifically against sight are coming onto the market as well. The company LE Systems has developed a flashing green light handheld laser, the Laser Dazzler,™ for distracting individuals.[60] It is supposed to render individuals sightless in daylight or at night but also be safe for eyes. Potential effects include the creation of fear, distraction and disorientation. In Figure 3.1 are indicated some of the deployment scenarios offered by LE Systems.[61] The US Department of Defense is considering mounting these on helicopters.[62]

FIGURE 3.1
PROPOSED SCENARIOS FOR LASER DAZZLING EQUIPMENT

Military Operations: In a sense America is constricted by the enormity of its power from using lethal force except in situations where a clear and unambiguous threat is presented. Terrorists can take advantage of this situation by hiding as harmless boat tenders servicing the fleet. When the rubber boat laden with explosives turned toward the *Cole*, only lethal options were available to the crew. Shoot to kill, and perhaps be wrong, or hesitate. Those were the options. The Laser Dazzler™ would have offered another option to the crew of the USS *Cole*. If the drivers of the rubber boat had been dazzled at 100 to 200 yards away, and still kept coming – even though the Laser Dazzler™ had taken away his vision, a hostile intent could have been inferred. Resort to lethal force, which the *Cole* had an ample supply, would have eliminated the threat ...

Correctional Facilities: Laser Dazzler™ is a perfect tool for cell extractions, depriving the prisoner of his eyesight while guards restrain him unseen. Response to a Prison riot situation benefits from an effective non-lethal projection of force. It matters not whether the laser beam lights up hostages or suspects, since it is ultimately harmless. Turn out the lights and laser down the entire population as authorities move in to re-establish control ...

Civilian Applications: Every high school in America should have a Laser Dazzler™ for a first response to a Columbine style incident.

Table 3.2 summarizes the main categories of non-lethal weapons. The previous description of the weaponry leaves a number of important developments uncommented upon. There has been a good deal of emphasis on delivery systems that cross the effect categories. Diehl Munitionssysteme of Germany, for instance, has reportedly developed a sophisticated multipurpose launcher with a range of 300 meters that can deliver payloads such as chemical irritants, marking paint, electrified or non-electrified nets and EMP equipment to jam communications equipment.[63] The US JNLW Directorate has funded the modification of unmanned aerial vehicles to dispense non-lethal payloads such as irritants, EMP devices and stingballs. Similarly, remotely controlled jet skis are being fitted with non-lethal capabilities.[64] It is not unheard-of that audio and visual stimuli techniques for 'mind control' make their way into considerations of this class of weapons.[65]

Besides electromagnetic pulse/microwave technology, there are other options under the heading of information warfare. These include communications-jamming equipment as well as more recently available

TABLE 3.2
TYPES OF NON-LETHAL WEAPONS

Category	Type	Description
Kinetic	Impact Munitions	Projectiles designed to incapacitate through inflicting energy to the body. Variations include plastic bullets, beanbags and sock rounds
	Water Cannons	High-pressured water stream
Mechanical	Entanglers	Nets and other ensnaring objects delivered by launchers, mines or other systems
	Barriers	Designed to stop individuals and vehicles
Chemical	Incapacitating Agents	Irritants and other agents (e.g., CS, CN, CR, OC) that are used as gases and sprays
	Receptor Chemicals	Meant to induce anxiety, submissiveness or fatigue
	Malodorants	Smells designed to induce particular emotional responses
	Aqueous Foams	Create a physical barrier and can be used with chemical irritants
	Superlubricants	Chemicals intended to reduce traction
	Superadhesives	Sticky materials for subduing individuals or denying access to facilities
	Supercorrosives	Chemicals introduced to degrade materials
Electroshock	Electroshock	Devices that deliver electrical shock, including projectiles, batons and belts
Acoustic	Infrasound Waves	Low frequency sound that allegedly causes disorientation and nausea
Lasers	Tactical Lasers	Discriminate and variable intensity lasers that heat and melt materials, including flesh
	Tagging Lasers	Meant to indicate the targeting of individuals
	Dazzling Lasers	Devices that produce temporary disorientation by overpowering visual senses
Optical Munitions	Visible Light Radiators	Devices that cause flash blinding
	Strobe Lights	High-intensity strobes meant to cause vertigo and disorientation
Microwave	Microwave Generators	Increase body temperature by heating water near skin

(Continued)

TABLE 3.2 CONTINUED

Category	Type	Description
Electro-magnetic (other)	Electromagnetic Energy	For disturbing or damaging communications and information equipment
Biological	Biodegrading Agents	Microorganisms for degrading materials and other functions
Other	Carbon Filaments	Tiny filaments designed to disrupt equipment
	Electronic Warfare	Means of disturbing or damaging communications and information equipment
	Psychological Warfare	Forms of (dis)information, such as radio broadcasts

techniques such as computer viruses and worms for wrecking financial, telecommunications and traffic-control systems without direct human fatalities. Psychological operations have also been discussed as a form of non-lethal warfare.

In recent years, with the establishment of designated programs of support, arguably something of a non-lethal 'bandwagon' has formed. Technologies such as passive-object sensors, sniper-detection systems[66] and swimmer- and diver-recognition devices have been forwarded as non-lethal weapons as they enable a greater selectivity in the use of force. Various pre-existing research agendas, such as the study of group animal behavior and human-animal communication,[67] have been billed as key in understanding non-lethal weapons. The example above of the airborne tactical laser provides an illustration of how weapons reported to enable precise targeting are thereby deemed non-lethal. Given the range of possible activities that can be included within a study of non-lethal weapons, it is important to clarify which topics will be considered in this book and why.

This analysis focuses on particular weapon technologies and related skills and practices. While a variety of actions undertaken in situations of conflict might not lead to injuries or physical damage (such as psychological operations and information warfare), it seems an unhelpful stretch of the term, however, to consider these as non-lethal *weapons*. Information and psychological warfare have long histories where there seems little reason to assume activities undertaken now or in the past have been done to incapacitate personnel or material while minimizing fatalities and undesired physical damage. The term *weapon* here is taken to mean material objects designed or used to inflict harm. Likewise, though sensor, detection and guidance technologies might figure as part of non-lethal weapon systems, they are only given consideration to the extent that they do so. Furthermore, in this book, the applications of weapons against humans rather than against equipment and materials are given the majority of attention. While the distinction between anti-personnel and anti-material applications is often difficult to make (anti-material weapons can affect humans and vice versa), the overall focus is with the former. The extent to which non-lethal weapons offer more legitimate options in the use of force hinges upon the possibilities of minimizing human injury and fatalities. Save for cases where humans are part of the systems targeted by anti-material weapons, it makes little sense to talk about anti-material weapons as 'non-lethal'.[68]

WE HAVE THE TECHNOLOGY?

The previous section adopted the rather conventional form of listing a range of technological options. Various weapons were described in terms of their stated characteristics and some key considerations. Following the general skeptical orientation outlined in the last chapter, though, it is worth stepping back from this analysis and considering the implications for the descriptions advanced.

Much of the current interest in non-lethals relates to so-called second-generation devices that have not yet materialized as working technologies. Limitations on what is publicly known about many non-lethals breed a good deal of speculation and conjecture about their feasibility and desirability. Jürgen Altmann suggests that this situation has fostered a series of unrealistic expectations. He has considered something of the basic plausibility of categories of non-lethal weapons and contended that many of the statements of their capabilities are, at best, immodest.

Take the case of acoustic weapons.[69] While a number of optimistic claims have been made in security and public publications about the potential of infrasound, upon closer examination such claims prove wanting for Altmann. Infrasound does not have the sort of drastic effects claimed for it at the intensity levels suggested by manufacturers and others. Certainly this type of sound can be annoying, but weapons based on it are not going to cause incapacitation, vomiting or defecation. The level required for such effects can only realistically be achieved in a sound chamber. While low-audio sounds (in the 50–100 Hertz frequencies) can produce intolerable responses, the necessary levels would almost certainly cause long-term hearing loss. The laws of nature and the ease of possible counter-measures (such as earplugs) mean acoustic weapons are of limited potential as non-lethal weapons. Many statements about the possibility of acoustic weapons were said by Altmann to be 'plainly untrue'.

Much the same has been said of other proposed weapons.[70] So, in relation to chemical corrosives, while the ability to dissolve tire fragments in laboratory conditions is easy enough, applying the necessary chemical over a whole airport is both impractical and of limited military advantage. The talk of biological microbes that turn fuel into jelly, or degrade metal, belies the conditions such as proper temperature, the availability of nutrients, and the creation of solutions for the microbes that are necessary for reactions to take place. Altmann suggests that sensationalist claims are often made in military and trade publications with little evidence to support them. Moreover, even those wishing to be critical of non-lethal weaponry have to rely on such statements as the basis for analysis. The lack of detailed technical information means commentators are often recycling claims that have little evidential basis.

As accounts of weapons become told and retold, the technologies can take on a definite status. During 2001, when police forces in Northern Ireland and mainland Britain were researching non-lethal weapons, a number of stories appeared in the British press about possible devices.[71] *The Observer* newspaper, for instance, reported on a 'Supergun' 'developed' in the United States that might be deployed in Britain during riots and periods of unrest.[72] This gun was reported as the 'holy grail' of policing, by combining electronic shock, pepper spray and video surveillance into one hand-held device. Sensors were said to judge the distance to targets, thereby enabling precise firing. The chemical and electroshock applications could be used together to reinforce effects. The video imagery could prevent unwarranted claims being made against the police regarding unnecessary use of force. It was reported that the supergun had 'received clearance from the US National Institute of Justice'. This article was based on a more extended one in the UK Police Federation magazine called *Police*.[73] That article gave a somewhat more qualified assessment of the device by noting that the National Institute of Justice had given clearance for initial trials pending the development of the first 'alpha' prototypes. However, further probing by the author found that, as of late 2001, the manufacturer of the gun was still seeking to raise money from investors to produce the device. Claims about the imminent uptake of the gun in Britain circulated, though.[74]

Given such considerations, the survey above – like other analyses – suffers from the basic flaw of taking statements made about weapons seriously when this might not be warranted. But to say that the technologies are not available is not meant to suggest that stories about them are therefore inconsequential. For Altmann, a major concern is that international treaties restricting the use of chemical and biological weapons in war might be weakened, given attempts to renegotiate them in order to allow the development of certain chemical and biological non-lethal weapons (see Chapter 7). If basic doubts can be raised about the practicality and feasibility of many second-generation weapons, then suggestions that treaties should be renegotiated or abandoned are ill-considered. The 'mere' reciting of previous descriptions about this technology is not unproblematic and not without its implications.

But I want to suggest here more than the importance of distinguishing 'fact' from 'fantasy' about non-lethals. In surveys such as the one given in this chapter, the conventions in language and forms of describing non-lethal weapons need to be examined. Brief descriptions of the origins, aims and effects of devices – say, kinetic-energy weapons being 'low-impact', 'non-lethal' or 'hazardous' – rely on background assumptions regarding the situation in which the technology would operate in practice. So, non-lethal compared to what, in what situations, and with what type of medical facilities at hand? Likewise, in giving an account of the aims associated with weapons, it is important to bear in mind that these aims are probably thought about differently by designers, policy-makers, users and recipients of technology.

Consider the points above in relation to questions about effects. The overview of non-lethals given so far has made various assertions about the stated and perhaps unintentional consequences of this technology. It is generally acknowledged that the effects of weapons can be complex and sometimes unpredictable. As will be argued in subsequent chapters, the effects of particular devices are often disputed and disputable. But a basic point is that statements about what particular technologies can and cannot do are the upshot of interpretations. To say, for instance, that pepper spray has or has not caused deaths raises important concerns about what one refers to by the term pepper spray, what situations are considered, with what degree of certainty claims are judged, who makes them and in relation to what supporting arguments. To make statements about 'pepper spray' at all is to imply that some definite thing exists which has certain implications. This is not to suggest that technologies do not exist, but our understanding of them is always formed through the interpretation of situations.[75] As will be illustrated in later chapters, in practice the general term 'pepper spray' stands for many devices that are taken by some to have varied characteristics. To simply say that certain chemical sprays have or have not caused deaths is somewhat problematic.

Reiterating points made in earlier chapters, given the alternative ways available for understanding technology, we need to consider how discussions of technology are presented. The issue at stake is more fundamental than simply calling for a qualification or further elaboration of statements. There are important questions about the orientation of analysts to claims advanced in conditions that are said to be characterized by uncertainty, disagreement and secrecy.

Take, for instance, the history of non-lethal weapons given above. It made a number of assertions about why devices were developed and thereby the wider context in which we should understand them. To begin a survey through a quote about the enforcement of 'law and order' in the colonies evokes a particular way of understanding the intent and likely applications of non-lethals. It will probably come as little surprise to most readers that keeping protests in the colonies down is not the historical background mentioned in sympathetic accounts of non-lethal weapons. Such an account links non-lethals to controversial contexts and applications of force that might taint assessments made. Likewise, those interested in promoting non-lethals often fail to make any links to the psychotropic drugs experiments of the 1950s and 1960s. Arguably, both these associations matter. The question is to what extent these backgrounds should inform evaluations made of the technology. A history told from the perspective of non-lethal weapons development in the Netherlands, where the requirements of recent peace-keeping missions have loomed large, would provide a much different context for understanding. The focus on the United States and Britain in this book is justified in that these countries are the sites of major efforts. They are also the countries most familiar to the author. Likewise, though, to concentrate on the anti-personnel applications of non-lethals is going to bring up the most controversial aspects of this technology.

Furthermore, this analysis has also implicitly adopted a certain 'bottom line' basis for the evaluation of the potential harmful effects of non-lethals. This has been one tied to a sanitized medical and technical language. To say that the immediate effects of the chemical irritants may cause the production of tears, coughing, sneezing, burning or even vomiting is a particular way of framing effects. In the actual situations of conflicts, the 'recipients', 'targets' or 'victims' might choose terms such as anger, humiliation, fear or racism to describe the effects of technology. Questions about the acceptability and legitimacy of weapons are ill-addressed without noting the competing framings possible. Again, the general point is that descriptions of technology are not merely just that, but are implicated in assessments made and how debates are characterized.

In short, the accounts of non-lethal weapons are open for questioning. What counts as realistic, fair and adequate accounts are key questions that need to be asked. The attempt will be made in the following chapters to acknowledge such considerations by reflecting on the basis of assessments and portrayals while making them. In this way, I seek to advance an understanding of the implications of the ambivalence of non-lethal technologies.

NOTES

1. See http://www.jnlwd.usmc.mil/default2.htm
2. Lewer, N. and Schofield, S., *Non-lethal Weapons* (London: Zed, 1997); Bunker, R. (ed.), *Nonlethal Weapons*, INSS Occasional Paper 15 (CO: USAF Acacdemy); Omega Foundation, *Crowd Control Technologies*, Report to the Scientific and Technological Options Assessment of the European Parliament, PE 168.394 (Luxembourg: European Parliament, 2000); and Selivanov, V., Klochikhin, V. and Pirumov, V., 'Modern Views on Development and Application of NLW in Anti-Terrorist and Peacekeeping Operations

(Summary of Russian-American Conference 1999, Easton, MD, USA)', in *Non-lethal Weapons*, Proceedings of the 1st European Symposium on Non-lethal Weapons, 25–26 September, Ettlingen, Germany (Postfach: ICT, 2001).

3. Jones, A. G., *Methods of Dealing with Civil Disturbances*, [PRO] *Secretary of State for the Colonies Circular – 1948* 14452/48, UK Public Records Office [PRO] CO 537/2712.
4. See Evans, R., *Gassed* (London: House of Stratus, 2000).
5. Mokalobe, M.,'No More Tears', *Track Two* 10, 3 (2001).
6. Robinson, J., 'Disabling Chemical Weapons', Presentation to PUGWASH Study Group on Implementation of the CBW Conventions, 27–29 May (Den Haag: PUGWASH, 1994) and Moreno, J., *Undue Risk* (New York: Freeman, 1999).
7. Evans, R., 'Drugged and Duped', *Guardian*, 12 March (2002), G4.
8. See Evans, *Gassed*, Chapter 8.
9. Ibid., p. 232.
10. Wisler, D. and Kriesi, H., 'Public Order, Protest Cycles, and Political Process', in D. Della Porta and H. Reiter (eds), *Policing Protest* (London and Minneapolis, MN: University of Minnesota Press, 1998).
11. Rodwell, R., 'How Dangerous is the Army's Squawk Box?', *New Scientist*, 27 September (1973), 730.
12. American System Corporation, *Joint Vision For Non-Lethals* (Dumfries, VA: American System Corporation, 1999).
13. Security Planning Corporation, *Nonlethal Weapons for Law Enforcement for the National Science Foundation* (Washington, DC: National Science Foundation, 1974).
14. Egnar, C., *Modelling for Less-lethal Chemical Devices* (Aberdeen: US Army Engineering Laboratory Technical Report, 1976).
15. Lewer and Schofield, *Non-Lethal Weapons*, Chapter 2.
16. Council on Foreign Relations, Independent Task Force, *Non-Lethal Technologies* (Washington, DC: Council on Foreign Relations, 1999).
17. See as well Alexander, J., 'An Overview of the Future of Non-Lethal Weapons', *Medicine, Conflict and Survival*, 17 (2001), 180–93.
18. Composed from the NIJ Online Portfolio at nij.ncjrs.org/portfolio/XAbout.asp as well as Altmann, J., 'Non-Lethal Weapons', *Medicine, Conflict, and Survival*, 17 (2001), 234–47.
19. See US/UK NLW Urban Operations Wargaming Program, *US/UK Non-Lethal Weapons/ Urab Operations Executive Seminar – Assessment Report*.
20. North Atlantic Treaty Organization, 'NATO Policy on Non-lethal Weapons', see http://www.nato.int/docu/pr/1999/p991013e.htm
21. Flatau, A., 'A Less-than-lethal Configuration for Delivery of Selected Chemical Agents', in *Proceedings of NDIA Non-Lethal Defense IV Conference*, 20–22 March 2000 (National Defense Industrial Association Conference, 2001) [cited 15 November 2001]. See www.dtic.mil/ndia/nld4/index.html
22. Harris, B., 1998, in *Proceedings of NDIA Non-Lethal Defense III Conference*, 25–26 February 1998 (National Defense Industrial Association Conference) [cited 15 November 2001]. See www.dtic.mil/ndia/nld3.html
23. See http://www.dtic.mil/ndia/nld4/buonodono.pdf
24. Arcidiacono, S., 'Recombinant Spider Silk Fibers', Presentation to the Non-Lethal Technology and Academic Research Symposium, 15–17 November (Portsmouth, NH: NTAR, 2000). See www.unh.edu/ntar/Transcripts/Finarcid.htm
25. See Joint Non-Lethal Weapons Program News, 'TIP Solicitation Completed', 2, 2 (1999), 4 – http://www.jnlwd.usmc.mil/default2.htm
26. Omega Foundation, *Crowd Control Technologies*, and The Sunshine Project, 2001, *Non-Lethal Weapons Research in the US: Calmatives and Malodorants*, Backgrounder Series #8 July.
27. Robinson, 'Disabling Chemical Weapons'.
28. Durant Y., Thiam, M., Petcu, C. and Vashista, N., 'Developing Microcapsules for NLW Applications', presentation to the Non-Lethal Technology and Academic Research Symposium, 15–17 November (Portsmouth, NH: NTAR, 2000). See http://www.unh.edu/ntar/PDF/Durant2.pdf
29. Gould, C. and Folb, P., 'The Role of Professionals in the South African Chemical and

Biological Warfare Programme', *Minerva*, XL, 1 (2002), 77–91.
30. The Sunshine Project, 2001, *Non-Lethal Weapons Research in the US*.
31. Pain, S., 'Stench Warfare', *New Scientist*, 7 July (2001), 42–5.
32. US Army and Monell Chemical Senses Center, *Establishing Odor Response Profiles*, Contract DAAD13-98-M-0064 April 1998.
33. Science Applications International Corporation, *Less-Than-Lethal Systems: Situational Control by Olfactory Stimuli*, June (San Diego, CA: SAIC, 1998).
34. See www.fas.org/faspir/pir0295.html for a version of this photo.
35. See Pilant, L., *Less-than-Lethal Weapons* (Washington, DC: International Association of Chiefs of Police, 1993).
36. Kittle, P., 'Aqueous Foam', in *Proceedings of NDIA Non-Lethal Defense IV Conference*, 20–22 March 2000 [cited 15 November 2001]. See www.dtic.mil/ndia/nld4/index.html
37. See Mathias, R., Mallow, W., Mason, R. and Collin, K., 'Non-lethal Applicants of Slippery Substances', in *Proceedings of NDIA Non-Lethal Defense IV Conference*, 20–22 March 2000 [cited 15 November 2001]. See www.dtic.mil/ndia/nld4/index.html
38. Collins, K. and Bowie, D., 'A History of Engine Defeat through Chemical Means', in *Proceedings of NDIA Non-Lethal Defense IV Conference*, 20–22 March 2000 (National Defense Industrial Association Conference) [cited 15 November 2001]. See http://www.dtic.mil/ndia/nld4/index.html
39. McNulty, J., 'Remote Controlled Non-Lethal Weapons Applications', in *Proceedings of NDIA Non-Lethal Defense IV Conference*, 20–22 March 2000 [cited 15 November 2001]. See www.dtic.mil/ndia/nld4/index.html
40. Barry, J. and Morganthau, T., 'Soon, "Phasers on Stun"', *Newsweek*, 7 February 1994, 26–8.
41. See Sunshine Project, *An Introduction to Biological Weapons, their Prohibition, and the Relationship to Biosafety* (Hamburg: The Sunshine Project, 2002) and Alexander, J., *Future War* (New York: St Martin's Press, 1999), Chapter 11.
42. Alexander, *Future War*.
43. For a general overview see Dando, M., *The New Biological Weapons* (London: Lynne Rienner, 2001), and for a technical analysis see Kagan, E., 'Bioregulators as Instruments of Terror', *Clinics in Laboratory Medicine*, 21, 3 (2001), 607–18.
44. Morales, F., 'Non-Lethal Weapons: Welcome to the Free World', *Covert Action*, 70, April–June (2001), 6–15; Council on Foreign Relations, *Non-Lethal Technologies*; Anon, 'Nonlethal Weapons give Peacekeepers Flexibility', *Aviation Week and Space Technology*, 7 December (1992), 50; Langford, D., *War in 2080* (London: Westbridge, 1979); and Pasternak, D., 'The Pentagon's Quest for Nonlethal Arms is Amazing. But is it Smart?', *US News and World Report*, 1997.
45. Scientific Applications and Research Associates, 'High Energy Toroidal Vortex for Overlapping Civilian Law Enforcement and Military Police Operations', in *Proceedings of NDIA Non-Lethal Defense III Conference*, 25–26 February 1998 (National Defense Industrial Association Conference) [cited 15 November 2001]. See www.dtic.mil/ndia/nld3.html
46. Sze, H., Gilman, C., Lyon, J., Naff, T., Pomeroy, S. and Shaw, R., 'An Acoustic Blaster Demonstration Program', in *Proceedings of NDIA Non-Lethal Defense III Conference*, 25–26 February 1998 (National Defense Industrial Association Conference) [cited 15 November 2001]. See www.dtic.mil/ndia/nld3.html
47. Deimling, L., Backhaus, J., Liehman, W. and Thiel, K., 'Infrapulse-Generator', in *Non-lethal Weapons*, Proceedings of the 1st European Symposium on Non-lethal Weapons, 25–26 September 2001, Ettlingen, Germany (Postfach: ICT, 2001).
48. Alexander, *Future War*, p. 65.
49. See e.g. Sample, I., 'Just a Normal Town ...', *New Scientist*, 1 July (2000), 18–24.
50. Wilbers, A., Naus, H., Vogten, T. and Zwamborn, A., 'Susceptibility of COTS PC to Microwave Illumination', in *Non-lethal Weapons*, Proceedings of the 1st European Symposium on Non-lethal Weapons, 25–26 September 2001, Ettlingen, Germany (Postfach: ICT, 2001).
51. See Walsh, N., 'People-Zapper Fires Microwaves at the Enemy', *Observer*, 18 March (2000).

52. Brinkley, M., 'The People Zapper', *Marine Corps Times*, 5 March (2001).
53. Ibid.
54. Alexander, J., 'Non-Lethal Weapons: The Generation after Next', in *Non-lethal Weapons*, Proceedings of the 1st European Symposium on Non-lethal Weapons, 25–26 September 2001, Ettlingen, Germany (Postfach: ICT, 2001).
55. Novichkov, N., 'Russia Plans To Export Non-Lethal Beam Weapon', *Jane's Defence Weekly*, 14 November (2001). Posted on the ArmsTradeList 13 November 2001.
56. See www.boeing.com/news/releases/archive1999.html
57. See http://www.nationaldefensemagazine.org/article.cfm?Id=744
58. See http://www.nationaldefensemagazine.org/article.cfm?Id=747 and Joint Non-Lethals Directorate, 'Pulsed Chemical Laser Proposal Selected for TIP Funding', *Joint Non-Lethals Directorate News*, 2, 1 (1998), 3.
59. JNLWD, 'USAF Laser Illuminator Programs', *Joint Non-Lethal Weapons Directorate Newsletter*, 3rd Quarter (2000).
60. LE Technologies, *The Laser Dazzler*, Handout produced for FPED III, 8–9 May (2001).
61. JNLWD, 'USAF Laser Illuminator Programs'.
62. Kauchak, M., 'Dazzled By the Light', *Armed Forces Journal*, July (2001), 20–1.
63. Sporer, M., 'Multipurpose Launcher for Non-Lethal Effectors', in *Non-lethal Weapons*, Proceedings of the 1st European Symposium on Non-lethal Weapons, 25–26 September 2001, Ettlingen, Germany (Postfach: ICT, 2001).
64. Alexander, 'Non-Lethal Weapons', in *Non-lethal Weapons*.
65. See e.g. Truesdell, A., *The Ethics of Non-lethal Weapons*, The Strategic and Combat Studies Institute, Paper Number 24 (Lancaster: Lancaster University, 1996) and Shukman, D., *The Sorcerer's Challenge* (London: Hodder & Stoughton, 1996).
66. Zierler, R., 'Passive Sensor System', in *Non-lethal Weapons*, Proceedings of the 1st European Symposium on Non-lethal Weapons, 25–26 September 2001, Ettlingen, Germany (Postfach: ICT, 2001).
67. See www.unh.edu/ntar/transcript.htm
68. In doing so, this analysis generally shares the conclusion of a joint UK/US conference about non-lethal weapons that 'the counter-materiel use of NLW is conceptually mean-ingless and the source of needless confusion'. See US/UK NLW Urban Operations Wargaming Program, *US/UK Non-Lethal Weapons/Urab Operations Executive Seminar – Assessment Report: 3*.
69. Altmann, J., *Acoustic Weapons* (Ithaca, NY: Cornell University Peace Studies Program, 1999).
70. Altmann, 'Non-Lethal Weapons', *Medicine, Conflict, and Survival*.
71. See e.g. Taylor, B., 'Poised to Arm our Police, the Star Trek Stun Gun', *Daily Mail*, 10 February (2001), and Walsh, N., 'People-Zapper Fires Microwaves at the Enemy', *The Observer*, 18 March (2001).
72. Bright, M., 'Riot Police to get US "Supergun"', *The Observer*, 3 June (2001).
73. Police Magazine, 'The Search for Policing's Holy Grail', *Police Magazine*, May (2001).
74. Diatribes, 'The A3P3 Super Gun', *Diatribes* [cited 28 November 2001]. See www.eight-ballmagazine.com/diatribes/diatribes001/diatribes14.htm
75. See Grint, K. and Woolgar, S., *The Machine at Work* (Cambridge: Polity, 1997).

4

Threats and Promises

Against what sort of threats are non-lethal weapons presented as effective and legitimate options? What critical points have been voiced? In supportive statements, non-lethals are said to provide a sliding scale of options that give greater flexibility in the use of force. They can thereby reduce the resort to lethal force and allow for more selective and appropriate responses. Such assessments contrast with more negative appraisals. The grounds for the latter include reasons as varied as undermining the effectiveness of security forces; making intervention more likely as it becomes less destructive; complementing and thus enhancing lethal force; creating unrealistic expectations about minimal bloodshed in conflict; causing unnecessary injury; and bringing a new arms race.[1]

As in previous chapters, the basic ambivalence of this technology is treated as stemming from the difficulties associated with classifying and evaluating a disorderly and diverse set of activities. Alternative evaluations at once hinge on relatively minor technical issues as well as major questions about the stability of the world order.

In acknowledging this ambivalence and outlining the promises and problems attributed to non-lethals, though, I want to avoid a simplistic treatment of debates. It would be easy to portray disputes in terms of clashes between anti- and pro-commentators. These could then be mapped on to different points of the political spectrum (left vs right)[2] or against specific organizations (civil liberty groups vs police and military forces).[3] Yet to portray things in this way already assumes much about why disagreement takes place and on what terms. In questioning such characterizations it is possible to acknowledge the diversity of assessments made, understand the negotiations and interpretations at work in evaluations, and acknowledge what is at stake in alternative accounts.

SECURITY CHALLENGES

As mentioned in the last chapter, with the end of the Cold War, alternative security threats have come to dominate the agendas of defence agencies in Western countries. For instance, in 1999, the Director of the US Joint Non-Lethal Weapons Office, Colonel George Fenton, argued that far from increasing international stability, the end of the Cold War has made the world a more volatile place.[4] The new security paradigm for Western countries, and the United States in particular, is based on threats from the proliferation of weapons of mass destruction, 'rogue states' (like North Korea, Libya and Iraq), regional instability and terrorism (particularly from Islamic countries). To these major threats can be added the growth of a broader spectrum of engagement of military forces in peacekeeping, non-combatant evacuations and humanitarian assistance missions. Demographic changes, including the worldwide growth in mega cities, are portrayed as placing further demands on military forces to be able to differentiate between combatants and non-combatants in theaters such as urban terrains. Fenton argued that the outcome of war in the future will not just turn on the destruction and annihilation of armies, but rather on the ability of force to be measured and legitimate.

One of the portrayed changes often alluded to in accounts of security challenges is the presence of diligent media and the related public unacceptability of death and injury. These place severe restrictions on acceptable actions.[5] The so-called 'CNN factor' means that real-time coverage of death and injuries inflicted by Western governments are aired around the world nearly instantaneously. Fears have been expressed that a hypersensitive public has made the forces of Western democracies quite risk-adverse. The ability of the military to wage war or the police to maintain order effectively is undermined by 'squeamishness' about killing or causing damage.[6]

The effects of the media extend into coverage of domestic policing as well. The filming of the beating of Rodney King in the United States in the early 1990s, for instance, is said to attest to the growing scrutiny placed on the police. While the increasing number of peacekeeping and other such operations by Western militaries has brought attention to the problems associated with the application of force in situations where civilians and combatants mingle, police forces have long had to act in settings where suspects and others mix. The unacceptability of death and injury from policing encounters are sometimes said to be increasing because of evolving legal restrictions and liability considerations. In 1985, for example, the US Supreme Court case *Tennessee v. Garner* ruled that deadly force was not acceptable against an unarmed, non-violent fleeing individual. This being so, alternative means of restraint and capture have been sought.

In some situations, the further deployment of non-lethals is set against the background of the general rise in assaults against the police, or the prevalence of handguns and other such weapons.[7] The latter is particularly true of countries such as the United States, where firearms are easily available to the general population. Officers need to respond to incidents in public order and routine policing in a manner that is reasonable for a given circumstance.[8] Situations might include violent persons intent on causing harm,[9] or armed but mentally ill individuals. Special Operations Commander Ijames[10] provides an illustration of the types of difficult situations in which force might need to be deployed:

> You are dispatched to a downtown business district on report of a disturbance: 'man armed with a machete'. Upon arrival, you find an elderly derelict slashing the air with a machete, in vain attempting to 'kill the demons'. He's screaming obscenities at passing pedestrians, and threatening to kill anyone who approaches him. What do you do?

TECHNOLOGIES OF GREAT EXPECTATION

The utility of non-lethals comes in finding options to deal with particular conflict situations and general large-scale security trends.[11] As two military analysts put it, 'The world is changing and our military role is changing. The tools [of the United States armed forces] don't seem to match the new roles we see out there. There is growing sense we need new tools.'[12] The basic philosophy said to guide these weapons is that death and injury are highly regrettable, so when it becomes necessary to employ force, non-lethals are supposed to minimize harmful effects. The diverse array of weapons available and being researched offer an intermediate stage of response between no force and deadly or serious force. In relation to military operations, the US Marine Corps states, 'A force armed with only traditional military weapons normally has only two options for effective compliance: maintaining a *presence* (essentially a threat) or actually employing deadly force. These two options are extremes with no middle ground.'[13] The Marines specify the benefits of non-lethals in these terms:

> Non-lethal Weapons expand the number of options available to commanders confronting situations in which the use of deadly force poses problems. They provide flexibility by allowing US forces to apply measured military force with reduced risk of serious noncombatant casualties, but still in such a manner as to provide force protection and effect

compliance. Because we can employ non-lethal weapons at a lower threshold of danger, commanders can respond to an evolving threat situation more rapidly. This allows US forces to retain the initiative and reduce their own vulnerability. Thus, a robust non-lethal capability will assist in bringing into balance the conflicting requirements of mission accomplishment, force protection, and safety of noncombatants. It will therefore enhance the utility and relevance of military force as a US policy option in an increasingly complex and chaotic international environment.[14]

Through such capabilities, non-lethals provide means of responding to the global security environment and the need to maintain public perceptions of the benign intentions of military interventions.[15] Military situations of potential use include crowd control, incapacitating individuals, clearing and denying access to areas and disabling equipment. A number of goals are said to guide the US JNLW Program and other such initiatives: enhancing operations, augmenting deadly force, allowing variable force capabilities, and increasing the overall acceptability of force.[16] Non-lethals can also act as so-called force multipliers, meaning that for any given situation, fewer personnel will be needed. Weapons are sought that are compatible with existing systems, that act nearly instantaneously, that ideally can be delivered by existing launchers, that are fairly simple to use and maintain, and that are 'tunable' in terms of their effects.[17]

Various scenarios have been offered for suggesting the potential of this weaponry. These extend from illustrations of how such devices can be used in particular tactical situations to how they might alter strategic considerations in the way conflict is fought. The RAND Corporation, for instance, has developed scenarios involving mega cities where attacks are conducted with a variety of non-lethal weapons combined with surveillance equipment.[18] Herein, remotely piloted vehicles could dispense chemical 'knock-out' agents over a given radius and individuals affected could be sorted through and divided between combatants and non-combatants.

Janet and Chris Morris have offered a vision of future warfare and the role of non-lethals therein:

> In fifty years, the battlefield is littered with wreckage of once-smart machines, now blinded, deaf, and dumb, and with human prisoners waiting to be collected by the U.S. and its allies …The urban landscape is dark, a victim of infrastructure attack from the air: Its power grids are off line, its

telecommunications and banking systems are dead; all media sources, now under U.S. control, advise surrender ... [The] U.S. Weapons of Mass Protection have disabled or incapacitated the enemy's war-making machines. Acoustic 'detonation waves' have driven off most of the enemy's warfighters. Air burst acoustic 'bullets' and electromagnetic rounds have disarmed, disabled, or routed the rest. Dense protein foams, some with CS additives, encase enemy fighting vehicles completely, blocking out vision and preventing the escape of those inside, who have barely enough oxygen to breathe.

... Electro-optical rounds have blinded enemy targeting systems. A few netted combatants struggle weakly, exhausted, stuck like flies to flypaper. Combined effect munitions (CEMS) have silenced anti-air batteries. Liquid metal embrittlement rounds have eroded metal weapons platforms so that occasionally big guns crumple of their own weight and wings fall from downed aircraft ...

On this 2050 battlefield, not only are human fatalities unacceptable to world public opinion, people have become ancillary to the issue of victory – or defeat. For the U.S. and its allies, high body count is no longer acceptable in pursuit of success in war.[19]

In law-enforcement applications, to non-lethal or less lethal weapons is generally attributed a more modest set of expectations. Writing in the early 1990s, sociologist P. A. J. Waddington[20] maintained that 'first generation' non-lethal weapons, such as irritant chemicals, water cannons and baton rounds, were fairly effective and appropriate options in arresting suspects, dispersing crowds or incapacitating individuals. This was especially the case in Britain, where a particular misplaced nostalgia existed within the police and elsewhere for 'traditional' forms of public-order policing involving static cordons of officers lined up against protesters. In the case of crowd-dispersal techniques, for instance, Waddington compares non-lethals with police techniques such as the baton charge. He maintains that rushing and striking crowds through the latter often resulted in indiscriminate effects, where those at the front of groups take the brunt of the injury whether or not they are the most violent individuals. As opposed to the baton charge, CS smoke can be employed in such a way as to compel a crowd to break up without serious injury and without necessarily targeting those individuals who happen to be at the front. Water cannons can be even more appropriate because they allow for the singling-out of individual troublemakers or ringleaders. Likewise in the context of rioting situations, devices such as baton rounds give the police the ability to incapacitate specific individuals. These technologies might cause pain but this should hardly be surprising as they are designed to be uncomfortable. The key point for Waddington is that non-lethal options provide the means for the more selective application of force than traditional methods and they typically have few lingering effects.

Drawing on domestic experiences in the United States and elsewhere, Jon Becker and Charles 'Sid' Heal[21] contend that non-lethals can be effective in riot-control situations. For them, understanding why requires considering something of riot dynamics. Riots are said to initially consist of crowds where individuals have varying degrees of commitment to being assembled. In such a setting, key individuals, such as agitators, often spark crowds in serious riots. Once a mob is created, individuals begin to lose their individual sense of identity. The longer a mob is formed, the worse the situation becomes. Therefore, equipment and tactics are needed that allow for an early intervention that can be directed against particular individuals. Less-lethal weapons permit such a response. Being affected by chemical irritants or kinetic weapons makes it 'personal' for individuals to stay in the fray. While non-lethals are not a panacea or a 'magic bullet', for Becker and Heal they do enable more appropriate force responses.

Public-order situations, be they riots or demonstrations, are not the only types of situation envisioned. Others include barricade and hostage situations, the break-up of domestic violence, the apprehension of fleeing felons, and the prevention of suicides with knives or guns. For those supportive of non-lethal weapons, among police departments thought of as progressive are those that seek to find novel options for dealing with these sorts of situation. 'Law-enforcement'-related situations also include incarceration and coast patrol. With regard to the latter, a member of the US Coast Guard maintains:

> From cocaine laden 'go-fast' racing across the Caribbean, to the rusted freighter loaded with hundreds of illegal migrants off the coast of California, to the foreign fishing vessels violating the US Exclusive Economic Zone, Coast Guard units are faced with the challenge of compelling compliance from individuals who have no desire to obey. The Coast Guard will continue its efforts to expand the use of non-lethal technologies to address all of these mission areas. Upon delivery of the JIP [Joint Integration Program] analysis

data on the riot control gear and large capacity OC pepper spray dispensers, the Coast Guard would like to produce dispensers for operational testing for use in migrant interdiction operations.[22]

Such is the varied potential accorded to non-lethal weapons.

REVOLUTION OR EVOLUTION

As indicated above and previously in this book, alternative promises are attributed to non-lethal weapons. On the one hand, more pragmatic, positive claims state that non-lethals will function as adjunct capabilities in warfare or policing. Such devices might cause less injury or prove more effective than current options for a particular situation. In this sort of assessment, the long history of such weapons is taken to suggest that their potential to drastically transform the use of force is unlikely.

Often, though, more visionary claims are offered. David Morehouse, author of one of the first books post-Cold War about non-lethals, claimed they offered humankind a revolutionary concept to escape from the growing lethality of war.[23] Individuals such as Janet and Chris Morris have echoed this sentiment by claiming that these technologies

> can paralyze without destroying an aggressor's infrastructure and military might. This is the face of war in the future: Conflicts must be settled by the world community while that community remains life-conserving, environmentally friendly, and fiscally responsible, even when faced with an enemy who is none of those things. Otherwise, the fielding and reconstruction costs of modern warfare will make war simply too expensive to fight, and victory impossible to achieve because of the inherent, unrecoverable economic loss of conflict itself.[24]

In policing situations, Sid Heal has argued that non-lethals are not 'simply kinder and gentler lethal weapons', but that between lethal and non-lethal options there are 'fundamental and distinctive differences in how they are defined and employed'.[25]

Often, these evolutionary and revolutionary claims mix quite readily. The US Joint Non-Lethal Weapons Program has so far focused its attention on fairly 'low-tech' or 'off-the-shelf' weapons, such as alternative methods of dissemination or variations of existing chemical and kinetic non-lethals. These are intended to serve as practical tactical aids.

Research into technologies designed to alter the manner in which war is fought has received less attention. Yet there is the stated desire to get beyond the tactical-level 'rubber bullet' type options.[26] New technologies are said to offer possibilities never afforded by past weapons. A US Joint Armed Forces paper on non-lethals offers the example of the *intifada* by Palestinians in the late 1980s as an instance of the limited applicability of previous response options.[27] While Israeli troops made use of some non-lethals, such as tear gas, '… the effects were limited by the low technology devices available, which proved inadequate to meet escalating civil unrest. When [Israeli] troops resorted to deadly force, the resulting civilian casualties undermined international support for the Israeli government policy.' In particular, the lack of 'stand-off' capability of past technologies, such as tear gas, has been cited as a major limitation.

Whether non-lethals are attributed with evolutionary or revolutionary potential is not simply a matter of semantics. These statements are important in determining how much emphasis should be placed on the overall class of technology and what expectations are considered realistic. So, are the likely advantages of the non-lethals going to be confined to a fairly narrow range of operational deployments or is the scope more widespread? If non-lethals are meant for incidents such as hostage-taking, few criticisms are likely to be voiced. Following the example given in the last chapter (see Figure 3.1, p. 51), though, the introduction of weapons as part of school security is likely to be seen as another matter. Claims about the novelty of current or future options also affect assessments about whether the limitations of past systems can be overcome. If some distinctive advantage is offered from a greater firing stand-off capability, then the problems associated with past technologies, such as tear gas, might be overcome. Conversely, if non-lethals are not a panacea, then the question should be asked as to whether they are being sold as such.

In keeping with the general orientation of this book, I don't wish to simply judge these different portrayals in favor of a particular side. Rather, the attempt is made to understand how revolution and evolutionary claims, as well as other types of claim regarding distinctiveness, are advanced and what these mean for the evaluation of the technology. In later chapters, such distinctions are examined in relation to questions of safety and testing considerations, the sorts of operational control believed to be appropriate, and the extent to which the pursuit of non-lethal weapons might mean a forgoing of international treaties. It will be argued that statements about the extent of the revolutionary potential of non-lethals vary depending on the contextual setting in question.

For the present time, though, let us move on to the doubts expressed about non-lethal weapons. In providing more suitable or flexible options, the justifications for non-lethals are meant to be quite persuasive. How can anyone reasonably deny police officers and armed-forces personnel the possibility of using non-lethal alternatives which cause less damage to personnel than that caused by present and/or previous means of enforcement?

ALTERNATIVE THREATS AND PROMISES

This section outlines the general grounds for criticism or words of warning. In doing so, it seeks to draw attention to the bases of alternative assessments. Many of the same characteristics forwarded in supportive statements provide the basis for conflicting evaluations.

A Precise Technology?

We can begin by noting that in many optimistic claims about non-lethals they are said to have precise effects of an understood and *relatively* harmless nature. To be as 'non-lethal' as possible, force must be able to be employed in a restricted manner and be fairly well known. Perhaps to a greater degree than most other technologies or types of weapon, non-lethals are discussed in term of their controlled application. Being relatively harmless often requires following certain proscriptions. That might relate, for instance, to the gap at which kinetic-energy weapons are fired or the amount of chemical exposure individuals experience. That crowds might disperse from settings because of the disorientating effects of a technology requires that there is a physical space into which to flee.

Uncertainty and unpredictability are all long-standing concerns regarding conflicts.[28] From a military perspective, Sheldon draws attention to these points and thereby calls into question the merits of non-lethals.[29] For him, the demanding and precise effects generally sought from these weapons are not possible, be that in warfare or peacekeeping. In such environments, these devices might add a degree of uncertainty in thinking what force ought to be deployed. In a similar vein, Robin Coupland of the International Committee of the Red Cross has questioned whether the prospects for temporary incapacitation are achievable in battlefield conditions.[30] In the messy conditions of warfare, expecting controlled force is said to be unrealistic. Moreover, even if some level of disablement were achieved, what would this mean for subsequent action? What, for instance, constitutes excessive force against a person 'known' to be temporarily 'disabled'? What additional actions might be necessary or justifiable to secure that person's quiescence?

The possibility for unanticipated consequences due to the unpredictability of situations extends far beyond questions about the effects of particular weapons in given situations. The contention that a specific or whole class of the technology could bring about alterations to the character of conflict has often proven wrong. During their introduction, the bow, artillery and tank were all said to be so effective that wars would come to a quick end.[31] These examples would suggest some caution in predicting the ultimate contribution made by particular technologies to how conflict is resolved.

Lethal or *Non-Lethal Force?*

In some supportive statements, non-lethals are justified as filling the large gulf between different force options or between deadly force and no action at all.[32] In justifying the intense development of non-lethal options, John Alexander has contended 'In my view, those who oppose the use of non-lethal weapons by fait must support either capitulation or killing.'[33] For him, the key question is 'Compared to What?'. Those opposed to non-lethal weapons must respond to this question. He further suggests that critics 'should ask of any "victims" they represent whether they would rather endure temporary pain or discomfort of non-lethal weapons – or be dead'.[34] In a somewhat similar vein, the US National Institutes of Justice, Science and Technology Division director, David Boyd, suggested that the '[p]olice still have the same choices Wyatt Earp had … they can talk a subject into cooperating, they can beat him into submission, or they can shoot him. What the police need are better alternatives.'[35] The wider the gulf presented, the greater the urgency and promise given to options said to span this chasm.

As previously mentioned, though, organizations such as the ICRC have contended that distinctions between lethal and non-lethal weapons are far from clear-cut. As such, framing the issues at stake in terms of doing something, doing nothing or causing death is unhelpful. Rather, it is prudent to determine and attend to the context-dependent and independent factors that influence the lethality of weapons. So, tear gas might be used to cause more deaths than would otherwise be the case. Conventional explosives, such as bombs, might be dropped near facilities to intimidate and frighten rather than kill. Posturing is more effective than is often portrayed. There are, in other words, various operational factors that enter into determinations of the effects of weapons, and these mean that sharp dichotomies can only cloud thought. Likewise, to say that the police today have the same options as Wyatt Earp is to downplay the role of tactics, intelligence, use-of-force training and conflict-management strategies for police-public encounters.

Further questions can be asked about the comparative grounds being made or assumed. For instance, the NATO Policy on Non-Lethal Weapons states that while the complete avoidance of fatalities or permanent injuries from non-lethals 'is not guaranteed or expected, Non-Lethal Weapons should significantly reduce such effects when compared with the employment of conventional lethal weapons under the same circumstances'.[36] Any comparison of the use of non-lethal weapons versus deadly force though presumes the latter would be legitimate recourse in particular situations.

The potential complementary use of non-lethal with conventional forms of weaponry has been a source of particular apprehension in critical evaluations. Quoting military and policy strategists, in 1994, Steve Aftergood argued that non-lethals would generally function as adjuncts to conventional weapons, thereby being more pre- than non-lethal. As such, the expression 'non-lethal' was 'politically attractive but purposively misleading'.[37] Here, the proper framing of discussions is not in terms of either non-lethal or lethal weapons but of both, and therefore more injury and death.[38] High-tech foams might lock people in a position or reduce their mobility only so that conventional weapons prove more deadly. The potential for non-lethals to enhance force in warfare is fairly widely acknowledged and even promoted.[39] Beyond these augmenting functions, novel types of possibilities might be created. The desire by the head of the US Joint Non-Lethal Weapons Directorate for 'a magic dust that would put everyone in a building to sleep, combatants and non-combatants'[40] is understandable from one point of view. Yet the development of calmative, sleep-inducing agents introduces the prospect of systematic rape in conflict. In short, weapons legitimized as a means of reducing injury might end up increasing it.

A Questionable Technology?

Calls in support of non-lethals are even made when things go wrong with the employment of such technology. As one prominent US commentator said in justifying the need for further funding:

> [The Rodney King incident] left Los Angeles reeling. In its aftermath, a commission was appointed, articles were written and debates ensued on the causes of such an incident and what could be done to prevent a recurrence. The media nearly bled its [sic] inkwells dry. Finding suitable non-lethal force tools for the police became a priority. When federal law enforcement agencies met the Branch Davidians outside Waco, Texas, the pressure increased even more, with

Attorney General Janet Reno calling for accelerated efforts
and additional funding to find tools that would subdue criminals without using deadly force.[41]

Such sentiments have been expressed elsewhere by senior US officials.[42]
That non-lethal weapons were central to the unfortunate outcome of
both incidents (King had been electroshocked as well as being blud-
geoned, and various chemical and acoustic-harassment non-lethal
weapons were used to deleterious effect at Waco) rarely tempers enthusiasm.

Such is the power of the abstract rationale given for non-lethal
weapons that they might have utility in any circumstance. When an
intervention goes wrong with traditional 'lethal' options, calls can be
made for new 'non-lethal' technology. Even after certain non-lethal
options are in place there is always the possibility of some marginal
benefit from additional technology. CS incapacitant sprays were introduced for routine policing in many British forces during the late 1990s.
Although justified as a means of filling a gap in the use of force, they
were later seen as insufficient. As a British Home Office official said in
relation to new research on a range of further non-lethals options,
'[b]etween CS spray and firearms there's a huge gap in the operational
requirements for the police. It would be enormously helpful for the
protection of officers if we could find a device to fill this gap ...'.[43] Even
if such statements and the general line of reasoning described are held
to be reasonable, whether it makes sense in terms of resource allocation
to continue to pursue technological options at the expense of other
priorities (say training or conflict management) is questionable.

Proceeding Adrift?

The potential for non-lethal weapons to complicate or confuse decision-
making about priorities and directions has been advanced along
different lines. Writing in the late 1970s in the middle of significant
social strife in Britain, the Council for Science and Society sought to
draw attention to possible future problems associated with the adoption
of various policing paraphernalia designed to be 'harmless'.[44] In particular, two worries were identified: technological drift and decision drift.
The former referred to the potential for substantial change in weapons
and techniques that could be brought over time because of constant
incremental innovations and alterations. The latter referred to the prospect for the gradual growth in the scope of deployment of a weapon. In
both processes, steady and relatively unnoticed changes could have
quite fundamental implications. For instance, a weapon originally intro-

duced as a last-step measure in a narrow range of situations, designed to achieve some specific desirable end (say officer or public protection), could, over time, become much more widely used, perhaps in ways never originally envisioned. While at each point the decision taken might be done for some reason, in the end, the overall cumulative developments could be highly undesirable. The specific concern expressed by the Council for Science and Society was the (para)militarization of the British police.

The potential for widening the scope of legitimate targets is illustrated in the quote given in the last chapter by the manufacturers of the laser dazzling device who suggested that in prisons, entire groups of individuals could be dazzled because of the ultimate harmlessness of such acts. How far such a philosophy is taken and who is targeted with what are matters of concern. In a related way, organizations such as the ICRC are concerned that civilians might gradually become treated as legitimate targets of attack in situations of conflict – as in the 'magic dust' quote on page 72 – because of the 'non-lethality' of certain weapons.

Other hazards related to the theme of 'drift' have been offered. 'Low-cost' decisions about intervention can lead to a gradual scaling up of conflict or resort to force. This danger is a variant of the well-known slippery slope argument.[45] In relation to military deployments this concept is sometimes tied up with that of 'mission creep'. Mission creep refers to a gradual shift in the objectives of forces. Alexander details such a process at work during operation *Restore Hope* in Somalia in the early 1990s. As told, while US forces went there to prevent a humanitarian catastrophe, the political state of near-anarchy in Somalia and the hostility of elements to outsider involvement meant that the mission moved from assistance to peacekeeping or even 'nation building'. Units such as the US Rangers were forced to counter threats from hostile locals. To do this, various raids were conducted and combatants pursued. For Alexander, 'it was a classic example of the wrong unit with the wrong equipment. The fault was not the Rangers, but that of senior officials who sent them to Somalia and then failed to provide adequate support.'[46] When a highly limited range of non-lethals, along with more appropriate tactics and training, were introduced in the *United Shield* withdrawal from Somalia, the operation proved more successful.

The force of the critical rub with drift, though, comes in acknowledging that having non-lethal force options and choosing *not* to act with them also poses certain dangers. If there are means of disabling equipment and personnel and they are not used, this in itself expresses a certain commitment, or at least has follow-on repercussions. For instance, if the United States had had viable non-lethal technologies for disabling equipment before Iraqi troops crossed the border into Kuwait in 1990, and decided not to use them at the time, then the appropriateness of subsequent large-scale military action might have been called into question. Taking the argument a step further, this situation becomes even worse if the capability to respond is thought to exist generally but such an estimate is based on unrealistic expectations of what technology can do.

To speak about decisions as adrift or unpredictable, though, implies a certain determination of the motivations and competencies of those in decision-making positions. So, choices about technologies are not being taken with some specific end in mind (such as the widespread uptake of a weapon) and their ultimate implications are not possible for individuals to grasp. In case of question of drift, whether or not one maintains this position hinges on the balance between seeing human action as controllable versus seeing it as uncertain and unpredictable. It is important to note that statements to the effect that particular results of action were foreseeable, planned, unintended or otherwise also have implications for suggesting whom to blame or commend. These issues will be taken up in later chapters.

An Escalatory Force?

Debates about the escalation potential of this weaponry extend from many of the preceding points. As noted above, one advantage cited with non-lethals is that they enable early intervention into conflict situations and are thus able to resolve situations before they become too serious. Some have gone so far as to suggest 'by the very nature of their purpose non-lethal weapons are non-escalatory'.[47] An opposed concern is that intervention might 'ratchet up' conflict. Some in the police, for instance, recognize that dangers for provoking crowds are associated with donning riot gear and deploying specific weapons. Drawing on internal British police documentation and interviews, Northam[48] cited a variety of negative implications attributed to using riot-control equipment. So, while protection shields might reduce injury, they were seen as likely to encourage people to throw objects at officers. Worries also existed that baton rounds or chemical irritants threatened to create a dependency on such instruments in the future.

Related concerns have been expressed about the dangers of escalation in war. Commenting on the history of 'non-lethal' chemical agents in warfare, the editors of the *Chemical Weapons Convention Bulletin* argued:

[Non-lethal weapons] extensively used in war include ethyl bromoacetate and congeners in the first World War; agent CN in Ethiopia (from December 1935), China (from late 1937) and the Yemen (1963); and agent CS in the Vietnam War and the Iraq–Iran war. In each case, these agents were used mainly or entirely not to avoid the use of conventional firepower but in conjunction with it, as a force multiplier. Moreover, starting in World War I, combat use of such gases preceded every significant outbreak of lethal chemical warfare.[49]

Many supportive claims about non-lethals typically stress that whatever the harm caused, they must be evaluated in relation to other force options. Here, the attempt is made to limit the range of issues at stake to those associated with a comparative analysis of a weapon's direct effects. Others have attempted to let the story run on a bit longer. For John Alexander, non-lethal weapons might stop dangerous cycles of violence from being reproduced. As he argues,

Most Americans were shocked to learn about the involvement of US Army troops in the My Lai Massacre. In that incident, young American soldiers, mostly draftees, willfully executed 347 unarmed men, women, and children. Herded into the ditch by the infantry platoon commanded by Lieutenant William Calley, the villagers were machine-gunned to death at point-blank range. The reason? Some of the soldiers' buddies had recently been killed in fighting in that area. Professional soldiers or not, combat evokes passion. Passion can get out of control, and frequently does. Retribution seems justified. The cycle continues.[50]

Here, an acknowledgement of the possibility for the unauthorized and tragic use of force prompts calls for new options. In a similar fashion, Chris and Janet Morris claim that non-lethal weapons will lead to a gradual reduction of levels of acceptable violence because these technologies will make killing less and less acceptable over time.[51]

Steve Wright takes the long view into account but offers a rather different assessment.[52] Examining the deployment of various forms of non-lethal weapons in Northern Ireland during protests and riots in the 1970s and 1980s, he argues that the increasing deployment and use of such technologies bred the very dissent they were designed to 'fix'. Water cannons, kinetic projectiles and tear gas were deployed in the context of a phased set of counter-insurgency tactics which meant that, as individuals began to find ways of subverting the effects of particular weapons, new and more violent tools had to be developed. So, while devices were meant to provide flexible responses, they locked the parties involved into cycles of violent conflict. The potential for non-lethals to lower the acceptability of death is also countered by the so-called 'brutalization effect'.[53] To the extent that non-lethals expand the range of circumstances where force is presented as a legitimate way of resolving disputes, the possibility exists that violence will be interpreted by some as a legitimate means of achieving other ends. The reduction of the acceptability of injury might be bought at the expense of making the resort to force more widespread.

A Finger in the Dike?

As outlined above, for proponents, a good deal of attention has been given to how non-lethals enable a response to a wide range of world security threats. Peace researcher Paul Rogers offers a rather different assessment of the potential of non-lethals to address major security challenges.[54] He does so by situating the types of global security concerns identified above in relation to a number of 'root causes'. These include resource constraints which mean that US and European standards of living cannot be universalized, and a growing division between a small, rich minority and the poor majority. In the future, major conflicts are likely to take place around issues such as migratory movements into countries of the North; resources conflict; anti-elite insurgencies that might take the form of religious conflict; as well as a host of rebellions and other conflicts which are impossible to predict. While these insecurities might lead some to respond with calls for non-lethals to provide more acceptable forms of conflict control, such efforts for Rogers amount to a 'thoroughly misguided strategy'. These weapons offer little more than stop-gap measures that might be employed by insurgents of various sorts. The use of non-lethals will fail to address the causes of insecurity and anti-elite actions. Thus, accounts of non-lethal weapons geared toward enhancing their potential to handle particular situations of conflict fail to give an appropriate description of the important features at stake.

A Technology for Prime Time?

The intense media scrutiny of death and injury caused by security forces is often cited as one of the central motivators for the development of alternative weapons. What is seen as a disproportional coverage to such incidents when done by Western militaries or police units necessitates even more caution than is currently given to the use of force. Alternative readings of the selective implications of media coverage, though, offer different ways of how the disparity between appearance and reality drives particular appraisals. In this regard, the names given to weapons, like the term 'non-lethal' itself, belie the actual severity of effects. A certain Orwellian 'Newspeak' characterizes the language of officialdom. Talk of 'rubber' bullets, 'tear' gas, 'stun' guns, 'beanbags' and the like might make these weapons seem relatively innocuous, but this is misleading. Such devices can cause death, vomiting, defecation, and traumatic internal injury. To equate the reaction to pepper sprays with eating 'a plate of fiery Mexican food'[55] is to suggest a misleading comparison.

Beyond just the names, though, many non-lethals are said to work in such a way as to make it impossible for those merely viewing their use to gauge the amount of pain inflicted. Some technologies, like tear gas, might be easily visible, but the intensity of its effects are not. Other instruments, like the proposed microwave Vehicle Mounted Active Denial System, would not have visible effects in any sense. It is this lack of visibility of effects, then, that is key in making non-lethals more 'CNN-acceptable' than many conventional weapons. While it is urgent that the weapons should *appear* safe, whether they *are* actually safe and the pain induced are different matters.[56] Along these lines, once this is acknowledged, it is possible to gain a greater understanding of the motives and intended effects of this technology. In the mid-1970s, the British Society for Social Responsibility in Science (BSSRS) advanced a number of these points in stating:

> The rubber bullet, for instance, created specifically for the Northern Ireland conflict, was designed to wound but not to kill, and to sound much more innocuous than it really is. The aim of these technologies is not primarily the physical elimination of opponents. Their target is the thoughts of opponents and potential opponents as much as their bodies.[57]

Likewise, in commenting on the widespread deployment of tear gas in Northern Ireland, including against those merely residing next to areas where there is a disturbance:

> ... Serves as a means of collective punishment for *all* the people in an area where political activity, whether 'violent' or otherwise, is taking place. The goal is to make the community politically ineffective by inducing them to withdraw support of the activities. Gas is very useful for producing

demoralization, because, as we shall show later, it singles out
for its worst effects the weakest members of the community.[58]

A specific fear of the BSSRS was that the principle of minimum political reaction rather than minimum force drove the development of the
weapons and their deployment in places such as Northern Ireland. This
public-relations function of non-lethal weapons has been made in many
other cases. As Rejali commented, 'We all remember how badly Rodney
King was beaten by L.A. police but no one remembers how many times
King was shocked and how much voltage he received.'[59]
    A general follow-on point is that historical and cultural processes are
at work in determining acceptability rather than the safety of scientific
gadgetry. What force is justified and to whom depends on who is
targeted. Whether non-lethals are used in conventional warfare, international peacekeeping or domestic policing bears on assessments of
appropriateness of force. While different standards of acceptability
apply between wartime combatants and non-violent protesters, whether
and what differences in treatment should pertain to, say, prisoners and
other civilians would appear to be a matter of contention.[60] As with the
BSSRS, Morales says that non-lethals might be made to appear benign
but that in truth this is far from the case.[61] The introduction of non-
lethals into the control of public protest, for instance, opens up the
potential for social control and social domination by 'Pentagon Inc.' as
civilians become presented as legitimate targets.
    In turn, P. A. J. Waddington has characterized some of these claims
about the difference between appearance and reality as misplaced. For
him, given that the police and others need ways of managing conflict
situations, then, what is needed are weapons that induce fear without
causing serious injury. 'Targeting the minds' rather than the bodies is
thus the most appropriate aim. He points to some of the limitations
associated with taking up the most acceptable options because of the
questions about public acceptability. The sjambok (a mix between a
heavy whip and a large stick), for instance, is seen as less injurious than
the standard police batons. The association of this device with the
Apartheid system in South Africa, though, means that it is not taken up
in Western countries despite its relatively mild effects.[62]

United We Stand?

One of the major threats accorded to certain non-lethals, particularly
biological and chemical weapons, is that their continued deployment
might undermine key international agreements. More or less space is
seen as provided for the development of certain such technologies by

such agreements. The concern is that under the banner of seeking weapons with the stated intent of minimizing harm, major agreements might be undermined or discarded. The possibilities surrounding this are discussed in Chapter 7.

Serving the World's Needs?

Another major area of criticism relates to the proliferation of non-lethals. As new ways of disabling, disorientating and dazzling become commercialized, the potential for them to spread out from their country of origin becomes ever more likely. While much of the current interest and funding of non-lethals takes place in the United States and Europe, there is potential for such technologies to become used outside of these places. Even if one assumes that non-lethals are valuable tools that are employed in proscribed ways for noble ends within Western countries, with their proliferation, the validity of positive assessments arguably becomes somewhat more questionable. Police and military forces are not a social good all around the world. The ability of non-state actors to obtain weapons also poses major concerns about how they might be employed, for instance, in terrorist acts. The availability of blinding laser weapons served as an early example of the types of fears expressed regarding the terror potential of non-lethals. While some individuals might respond to the events of 11 September 2001 by redoubling efforts to develop an ever-wider range of ways of targeting the body, others would respond with caution to inventing new weaponry.

GROUNDS FOR DIFFERENCE

From the previous section it is apparent that divergent, sometimes diametrically opposed, views are given of non-lethal weapons. Even in relation to the same set of concerns, supportive and critical statements differ markedly. Whether non-lethals might escalate levels of conflict depends on what one takes as the appropriate frame of reference for addressing this issue. At once, statements about non-lethals are attempts to speak about the intent of those promoting them, the relevant contexts for their consideration, the competencies of those managing them and an assortment of other factors. How one defines the possible disjuncture between the appearance and the reality of weapons and what this means for their public-relations potential speaks to a wide variety of considerations.

It is not just that opposing views provide alternative ways of making sense of the same technology, but different versions of what non-lethals can and cannot do are on offer. In this sense, some divergent views about the weapons are not easily resolvable: say through the provision of more technical data. Such different ways of making sense of situations have important implications.

Take the matter of how non-lethals function. In many statements, they are considered simply as tools that allow service personnel or officers to achieve some given end. Herein, as discussed in Chapter 2, their effects depend on the choices made by those using them. 'Intent', in other words, 'is everything'. Critics of non-lethal weapons 'misconstrue technology with human intent'.[63] Given that these weapons are explicitly justified and deployed to be less deadly than conventional ones, if there is any blame to be cast it typically rests with those in positions of authority who make major deployment decisions. As Alexander says, 'non-lethal weapons are designed with the intent of limiting physical damage. Nothing will prohibit misapplications or eliminate accidents, which are training and management issues.'[64] To say that non-lethals might bring a premature use of force, for instance, is to blame technology for faulty human choices. Instead, new technologies with less lethal effects could result in less loss of life and therefore less desire for retribution. Members of the BSSRS itself shared this framing of the legitimacy of 'riot-control' technology in terms of its intent. They took that intent, though, to stem from a desire of crises-ridden capitalist governments and their security apparatuses to suppress dissent.

Whatever the intent in question, whether that be benign or malicious, the suggestion in this argument is that the technology is itself not a source of concern. Non-lethals might enable lives to be saved in some situations, but in others a different outcome will emerge. Thus the argument is a moralist one – technology *adds* capabilities to individuals whose motives are either correctly or incorrectly placed.[65] This line of reasoning has obvious importance for the allocation of responsibility and blame. As non-lethals are designed in such a way as to minimize injury, then the range of concerns for thinking about why there might be unfortunate outcomes center on malicious intent, incompetence, or acknowledging certain unintended side effects.

Other characterizations have been offered to make sense of how non-lethals function. As noted above, both within and outside security agencies, suggestions have been made that the availability of non-lethal weapons *transforms* the options and likely actions of users and organizations. Typically, the change brought about is due to the technology operating in wider environments, rather than because of the device itself causing certain outcomes. Just what sort of transformation is initiated differs in these accounts. That might be, to some degree, encouraging the resorting to force and the fixing-in of violent patterns of behavior. Such contentions are corroborated by general research into the so-called 'weapons effect'. Psychologists have argued that the presence of weapons tends to 'prime' aggressive behavior in individuals.[66] In laboratory and field settings, the mere presence of weapons influences both the interpretation and assessment of the proper response to a given situation.[67] The transformative ability of the availability of non-lethal weapons is not limited to negative implications. The contentions that such weapons are 'designed to keep people alive' or 'by the very nature of their purpose non-lethal weapons are non-escalatory' imply to some degree that the availability of the technology itself has implications for the management of conflict.

These multiple characteristics of how non-lethals function impact directly on what they are supposed to accomplish in practice. A key line of debate centers on whether these weapons merely accomplish a given task or whether their ultimate implications are more diverse. In general, many non-lethals are said to offer means of what might loosely be referred to as 'controlling' situations – dispersing crowds, incapacitating individuals, and rendering equipment unusable. The incorporation of non-lethals into security agreements might thereby have consequences for how conflicts are handled.

Arguably, the debate about the merely additive or transformative capabilities of non-lethal weapons has much in common with that about those for conventional guns. In debates about controls on handguns in the United States, the 'pro-gun' argument that 'guns don't kill people, people do' clashes with arguments about such weapons changing behavior. As Latour argues, such debates turn on whether or not the goals of individuals and the functions of the gun are mediated when brought together. In their vulgar forms, such debates suffer from the problem of starting with fixed essences of goals and functions. Discussions about deaths from gun use that try to pin down either the gun or the individual *alone* as the cause are bound to be problematic. In later chapters we will see, in controversies about the use of non-lethals, just how those commenting attempt to cut through assemblages of human and non-humans (or refrain from doing so) in relation to particular notions of 'the context' in order to allocate blame and praise.

FRAMINGS OF AMBIVALENCE

While the last chapter focused on claims made about various existing and proposed non-lethals, this one has taken as its main consideration the situations of their past and envisioned usage. Drawing on alternative contexts, background assumptions and implications, different sorts of assessments can be offered about the likely implications of non-lethal weapons. Whether, and in what manner, we understand a given weapon in relation to its immediate or long-term effects, its use within a fairly limited number of situational areas or its potential wide-scale deployment, and its likely visibility to, and given by, media organizations will inform the type of assessment made.

For now, two major implications will be followed through. First, once these points are acknowledged, it becomes easier to understand some of the more messy and inconsistent views expressed about this technology. The notion that sharply divided camps exist between anti- and pro-non-lethal weapons positions is misleading. Arguably, the situation is more complex and dynamic. This can be seen in a number of ways. As has already been mentioned in the case of blinding lasers, the basis for criticisms offered for or against certain weapons is not as one might expect. As we will see in following chapters, multiple positions are taken about the uptake of particular weapons, even within a certain organization.

Individual commentators as well, though, often hold more complicated positions than simply 'for' or 'against' particular activities. Although a long-time proponent of non-lethals in the 1980s and 1990s, P. A. J. Waddington cast some doubts on the wave of enthusiasm for non-lethals in Britain that developed during 2001. After a number of police shootings during that year, the potential of less-lethal weapons received quite a bit of media and police attention. As a counter to this, Waddington argued that non-lethals would not be used in most cases where armed officers were deployed because those officers are specifically called out in Britain when lethal force is required; deaths would inevitably take place with non-lethals; talk and negotiation were still the best options for officers; a tension exists between minimizing risks to officers and those to suspects; and the public will remain ignorant of good deployments of non-lethals and only become informed when unfortunate outcomes take place.[68] So controversies about the legitimacy of the use of force will remain, whatever the technological capabilities. The senior advisor for police weapons in Finland, Jorma Jussila, takes a somewhat similar line on the limitations of non-lethals in noting that whatever their operational utility, their public legitimacy is dependent on open and accurate police forces as well as the credibility of the control procedures in place.[69]

Critical evaluations of non-lethal weapons are subject to revision or qualification as well. John Alexander quotes a personal email from Steven Aftergood in 2000. Despite offering some of the earliest critiques of recent non-lethal initiatives (see earlier in this chapter, page 72), Aftergood writes:

83

> I was going to say that I have thought of you a couple of times over the past few weeks. In particular, I felt that I may have done a serious disservice in my criticism of non-lethal weapons. Watching the violence between Israelis and Palestinians unfold, with the accompanying deaths, I couldn't help but wish for the easy availability of non-lethal weapons. The largely hypothetical concerns I had suddenly seemed less compelling. Maybe it is not too late to do something about it.[70]

Other than such drastic shifts in thinking, there are few commenting on non-lethal weapons that could be said to be against them per se. While their employment in particular public-order situations might be of substantial worry, in other situations critical points would be more muted.

MOVING ON

The argument of this part of the book would suggest that a detailed analysis of the claims about this technology is necessary. Just where and how one draws lines of appropriateness are matters for close scrutiny. The scope given to the deployment of non-lethals, how they are said to function and by what measures they are evaluated are all matters of potential dispute. While it might be agreed by most that non-lethals are not a panacea, what would count as an overselling or mis-selling is not clear. Moreover, close scrutiny is needed because, even in the simplest accounts of weapons, there are important concerns about the framing of discussions. As we proceed, we need to think about the basis for the evaluations given and the limitations therein. Beyond just drawing attention to the conflicting accounts of weapons, it is important to consider just how such accounts are employed. It is necessary to move beyond general claims about the safety and merits of technology, to consider instead what claims are made in relation to specific cases. As will be argued, though, this movement into greater detail does not take us from the land of opinions to that of facts, but rather brings further questions about the interpretation of technology.

NOTES

1. For an overview, see Council on Foreign Relations, *Non-Lethal Technologies: Military Options and Implications, Report of an Independent Task Force* (New York: Council on Foreign Relations, 1995), and Lewer, N. and Schofield, S., *Non-Lethal Weapons* (London: Zed, 1997).
2. Morris, C. and Morris, J., 'Moving Beyond the Rubber Bullet', Presentation to Non-Lethal Technology and Academic Research Symposium, 15–17 November (Portsmouth, NH: NTAR 2000). See www.unh.edu/ntar/NTARII.htm
3. Heal, C., 'What will the "Magic Bullet" Look Like?', in *Non-lethal Weapons*, Proceedings of the 1st European Symposium on Non-lethal Weapons, 25–26 September 2001, Ettlingen, Germany (Postfach: ICT, 2001).
4. Fenton, Colonel, 'Fielding Non-Lethal Weapons in the New Millennium', Presentation at Jane's Non-Lethal Weapons Conference, London, 1–2 November (1999).
5. See Truesdell, A., *The Ethics of Non-lethal Weapons*, The Strategic and Combat Studies Institute, Paper Number 24 (Lancaster: Lancaster University, 1996), and Alexander, J., *Future War* (New York: St Martin's Press, 1999).
6. Shukman, D., *The Sorcerer's Challenge*.
7. Mangold, T., *Panorama – Lethal Forcee*, aired BBC, Sunday 9 December 2001.
8. Levenson, H., Fairweather, F. and Cape, E., *Police Powers* (London: Legal Action Group, 1996).
9. Meyer, G., 'Nonlethal Weapons vs. Conventional Police Tactics', *The Police Chief*, August (1992), 10–19.
10. Ijames, S., 'Less Lethal Force', *The Tactical Edge*, Summer (1995), 51–5.
11. Becker, J. and Heal, C., 'Less-than-Lethal Force', *Jane's International Defence Review*, 62–4.
12. Barry, J. and Morganthau, T., 'Soon, "Phasers on Stun"' *Newsweek*, 7 February 1994, 26–8.
13. Marine Corps, *Joint Concept for Non-Lethal Weapons* (Quantico, VA: Marine Corps, 1998). See www.hqmc.usmc.mil/nlw/nlw.nsf
14. Ibid.
15. Lovelace, D. and Metz, S., *Nonlethality and American Land Power* (Carlisle, PA: Strategic Studies Institute, US Army War College, 1998).
16. Steele, Lieutenant General M., 'Non-Lethal Weapons and the Warfighter', in *Proceedings of NDIA Non-Lethal Defense III Conference*, 25–26 February 1998 (National Defense Industrial Association Conference) [cited 15 November 2001]. See www.dtic.mil/ndia/nld3.html
17. Marine Corps, *Joint Concept for Non-Lethal Weapons*.
18. Glenn, R., 'Non-Lethal Weapons and Urban Operations', Presentation at Jane's Non-Lethal Weapons Conference, London, 1–2 November 1999.
19. Morris, J. and Morris, C., 'Tomorrow's Battlefield', Presentation to the Non-Lethal Technology and Academic Research Symposium, 15–17 November (Portsmouth, NH: NTAR, 2000). See www.unh.edu/orps/nonlethality/pub/tomorrow.html
20. Waddington, P. A. J., *The Strong Arm of the Law* (Oxford: Clarendon, 1991), Chapter 6.
21. Becker, J. and Heal, C., 'Nonlethal Weapons and Peacekeeping Riot Control', www.nonlethal.com/reference/articles/bh01.htm
22. Jacobs, B., 'Expanding the Continuum of Force', *Joint Non-Lethal Weapons Directorate Newsletter*, 4th Quarter (2000), 2–3.
23. Morehouse, D., *Nonlethal Weapons* (Westport, CT: Praeger, 1996).
24. Morris and Morris, 'Tomorrow's Battlefield'.
25. Heal, 'What will the "Magic Bullet" Look Like?', in *Non-lethal Weapons*, 15–2.
26. Fenton, 'Fielding Non-Lethal Weapons in the New Millennium'.
27. Marine Corps, *Joint Concept for Non-Lethal Weapons*.
28. See Keegan, J., *The Face of Battle* (London: Pimlico, 1991).
29. Sheldon, Rex, *Nonlethal Policy, Nonlethal Weapons, and Complex Contingencies* (Newport, RI: Naval War College, 1999).
30. Coupland, R., ' "Non-Lethal" Weapons: Concerns of the International Committee of the

Red Cross', Presentation to The Subcommittee on Security and Disarmament, Committee on Foreign Affairs, Security and Defence Policy, European Parliament, 2 May 1998.

31. See Creveld, M. van, *Technology and War* (London: Free Press, 1991) and Dunlop, C., 'Technology: Recomplicating Moral Life for a Nation's Defenders', *Parameters*, Autumn (1999), 24–53.

32. Kehoe, J., 'Non-Lethal Laser Baton', in J. Alexander et al. (eds), *Security Systems and Nonlethal Technologies for Law Enforcement* (Bellingham: Washington, 1997).

33. Alexander, J., 'An Overview of the Future of Non-Lethal Weapons', *Medicine, Conflict and Survival*, 17, 3 (2001), 180–93.

34. Alexander, J., *Future War* (New York: St Martin's Press, 1999), 187.

35. Pilant, L., *Less-than-Lethal Weapons* (Washington, DC: International Association of Chiefs of Police, 1993).

36. North Atlantic Treaty Organization, *NATO Policy on Non-lethal Weapons*. See www.nato. int/docu/pr/1999/p991013e.htm

37. Aftergood, S., 'The Soft-Kill Fallacy', *The Bulletin of Atomic Scientists*, September/October 1994, 40–5.

38. For a number of accounts of non-lethals being used to complement conventional weapons see Omega Foundation, *Crowd Control Technologies – Technical Annex*, Report to the Scientific and Technological Options Assessment of the European Parliament, PE 168.394 (Luxembourg: European Parliament, 2000).

39. Lovelace and Metz, *Nonlethality and American Land Power*.

40. Edwards, R., 'War without Tears', *New Scientist*, 9 December (2000), 3. Quoted from Dando, M., 'Future Incapacitating Agents', in N. Lewer (ed.), *The Future of Non-Lethal Weapons* (London: Frank Cass, 2002).

41. Pilant, L., 'Adding Less-than-lethal Weapons to the Crime-fighting Arsenal', *The Journal*, Fall (1994) [cited 7 March 2001]. See www.zarc.com

42. Nollinger, M., 'Surrender or We'll Slime You', *Wired*, February (1995).

43. Bennetto, J., 'Police to Test Stun Guns and Weapons Firing "Spiderwebs"', *The Independent*, 10 February 2001, p. 6. This argument is somewhat inaccurate as most British police forces place the baton *after* CS sprays in their continuum of force and thus between the sprays and firearms.

44. Council for Science and Society, *Harmless Weapons* (London: Barry Rose, 1978).

45. Council on Foreign Relations, *Non-Lethal Technologies* (Washington, DC: Council on Foreign Relations, 1999).

46. Alexander, *Future War*, 23.

47. Kirkwood, Lt. Col. D., *Non-Lethal Weapons in Military Operations Other Than War* (Newport, RI: Naval War College, 1996).

48. Northam, G., *Shooting in the Dark* (London: Faber & Faber, 1988).

49. Chemical Weapons Convention Bulletin, 4 (March 1992), 15. Quoted in Harper, Major E., 'A Call for the Definition of Method of Warfare in Relation to the Chemical Weapons Convention', *Naval Law Review*, XLVIII (2001), 149.

50. Alexander, *Future War*, 15. There would be few that doubted the tragedy of such massacres and the need to prevent them. Yet how this might be done depends on how such incidents are understood. Social psychologists Herbert Kelman and Lee Hamilton, for instance, suggested that the My Lai Massacre is better thought of as a crime of obedience than as a crime of emotion. (See Kelman, H. and Hamilton, V., *Crime of Obedience* [New Haven and London: Yale University Press, 1989].) The exercise of authority was central to the incident and later evaluations of who was to blame. To elaborate slightly, the chain of command up from Second Lieutenant Calley was characterized by vagueness about what the troops were supposed to do that day. The soldiers were given orders to root out Viet Cong insurgents and there were not supposed to be any innocents in the village when the US forces came in. Calley claimed he had been given orders to kill on the day and he certainly gave orders to kill during the massacre. Despite others carrying out his commands, he was the only person charged or convicted of any crimes. Elements of the Army command, for their part, tried to cover-up the incident (see Hersh, S., *Cover-up* [New York: Random House, 1972]). Had a traveling army photographer not captured images of the mission it might well have gone down in history as a success.

The point of this example in relation to this book is that intent to kill in itself was not the principal dynamic. Instead they argue that it is 'more instructive to look not at the motives for violence but at the conditions under which the usual moral inhibitions against violence become weakened' (pp. 15–16). Kelman and Hamilton identify three such conditions: when acts are authorized and individuals do not feel personally responsible, when actions become so routine that moral questions are simply not raised, and when the recipients of acts become demonized. These processes perhaps help explain that, however shocked by the events, most Americans at the time said that they would have followed orders to kill and most even thought Calley should not be brought to trial for his actions. In relation to these sorts of consideration, the likely merits of non-lethals to avoid long-term cycles of violence become more disputable. To the extent that the 'non-lethal' effects of particular weapons help justify a greater scope for the application of force or enable a greater psychological distance between operators and recipients, they may make violence more acceptable.

51. Morris and Morris, 'Moving Beyond the Rubber Bullet'.
52. Wright, S., *New Police Technologies & Sub-State Conflict Control*, PhD Dissertation (Lancaster: University of Lancaster, 1987).
53. Melossi, D., 'The "Economy" of Illegalities', in D. Nelkin (ed.), *The Futures of Criminology* (London: Sage, 1995).
54. See Rogers, P., *Losing Control* (London: Pluto Press, 2000) and Rogers, P., 'Understanding the Adversary: The Evolution of Civil Unrest', Presentation at Jane's Non-Lethal Weapons Conference, London, 1–2 November 1999.
55. Pilant, *Less-than-Lethal Weapons*.
56. Wright, S., *An Appraisal of Technologies of Political Control*, Report to the Scientific and Technological Options Assessment of the European Parliament, PE 166.499 (Luxembourg: STOA, 1998).
57. Ackroyd, C., Margolis, K., Rosenhead, J. and Shallice, T., *The Technology of Political Control* (London: Pluto, 1980).
58. Ibid., 27.
59. Rejali, D., *Electric Torture*, February (1999). See internationalstudies.uchicago.edu/torture/abstracts/dariusrejali.html
60. For a discussion of prison standards and non-lethals see Amnesty International, *Cruelty in Control?* AMR 51/54/99 (London: Amnesty International, 1999).
61. Morales, F., 'Non-lethal Weapons: Welcome to the Free World', *Covert Action*, 70, April–June 2001, 6–15.
62. Waddington, P. A. J., *The Strong Arm of the Law* (Oxford: Clarendon, 1991), Chapter 6.
63. Alexander, 'An Overview', see note 33.
64. Alexander, *Future War*, 6.
65. This characterization of the debate stems from Latour, B., *Pandora's Hope* (Cambridge, MA: Harvard University Press, 1999), Chapter 6.
66. Berkowitz, L. and Le Page, A., 'Weapons as Aggression-Eliciting Stimuli', *Journal of Personality and Social Psychology*, 7 (1967), 202–7.
67. For a recent overview of the 'weapons effect', see Anderson, C., Benjamin, A. and Bartholow, B., 'Does The Gun Pull the Trigger?', *Psychological Science*, 9, 4 (1998), 308–14.
68. Waddington, P. A. J., 'Less-Lethal Means Less Effective', *Police Review*, 7 September 2001.
69. Jussila, J., 'Future Police Operations and Non-Lethal Weapons', *Medicine, Conflict and Survival*, 17 (2001), 248–59.
70. Alexander, 'An Overview'.

PART II:
TECHNOLOGIES, CONTEXTS AND CONTROLS

5

Weapons of Minimal Harm?: Assessing Effects

The chapters in this part of the book examine key aspects of non-lethal weapons: their effects (Chapter 5), employment (Chapter 6), and the prospects for their control (Chapter 7). As discussed previously, much (but by no means all) of the impetus for non-lethal weapons is said to derive from their ability to minimize injury. Their legitimacy, at least in part, depends on their harmful effects being deemed 'relatively acceptable'. Of course, the basic rub with such weapons is how they can simultaneously inflict enough energy or pain as to compel certain behavior while being relatively safe. Herein, the issues arise of what constitutes acceptable safety levels and how those ought to be determined. Even if one assumes that weapons should be considered 'non-lethal' by comparison to other use-of-force options, the issue must be addressed of what other options are to serve (and in what conditions) as the basis for comparison. If it is assumed that conventional weapons, such as firearms, are 100 per cent lethal and they are the standard for comparison, then the evaluation of the weapons discussed in Chapter 3 becomes predictable, but arguably questionable as well.

While the preceding chapters have discussed effects in general terms, this one begins a more detailed analysis. It considers the role of scientific and technical expertise in assessing just how safe is 'relatively safe'. Following the points made at the end of the last chapter, the attempt is made here to get beyond simple characterizations of the debate about non-lethal weapons in order to consider the process for identifying, interpreting, evaluating and managing risks. Chapter 2 suggested that much goes on behind claims and figures about injuries from non-lethals. 'Causation' in relation to injuries, for instance, is often a slippery notion. As in other areas of scientific debate and technological assessment,[1] it is necessary to consider how 'facts' and criteria for assessment are themselves bound up with the interpretation of evidence. Determinations of effects are the outcome of negotiations between security officers, medical practitioners, policy-makers, regulators, journalists and others. To the extent that consensus exists about the safety or non-safety of particular weapons, this is taken as an achievement based on the interpretation of evidence.

While this chapter cannot hope to provide a comprehensive treatment of all non-lethals, by examining two areas, I hope to illustrate the types of uncertainty and dispute in evaluations. The evidence and procedures for assessing the effects of particular weapons are considered with a view to asking how safety and risk claims are secured and questioned. In taking a skeptical approach to statements about non-lethals, I give particular attention to the claims of those who are most often, and arguably most able, to offer assessments: manufacturers and security agencies. That many in these organizations offer supportive assessments means that much of the following argument is geared toward unpacking optimistic claims. Questions and tensions are discussed throughout, though, about positive and negative assessments.

ASSESSING NON-LETHAL WEAPONS

By way of beginning an analysis of the safety claims and risks, it is worth making a few preliminary points generally widely recognized as factors determining the safety and risk. At a basic level, the effects of particular weapons are conditional on the manner of their employment. In most cases, the amount of force, its duration, over what distance it is applied, and with what sort of accuracy all contribute to defining a weapon's lethality. If the only thing that separates a medical drug from a poison is dosage,[2] then even the most benign of substances might cause severe injury. Where and how the line should be drawn between acceptable and unacceptable effects are areas for discussion.

Complicating the picture is the variability of effects often said to exist between individuals. A particular riot-control agent, for instance, might have little effect on some individuals but result in substantial injury for others. Reasons for this might include weight and constitution variations, pre-existing conditions such as asthma or heart disease, or the consumption of prescribed or illicit drugs.[3] The mechanisms by which particular weapons act may be more or less understood; for instance, more is known about the mechanisms of kinetic impacts than electrical discharges.

As an illustration of the types of difficulty that might attend upon non-lethal weapons, take the case of aqueous foam mentioned in Chapter 3. The US National Institute of Justice has funded research into such foams as a response to individual or group disturbances in prisons.[4] The basic idea is that belligerent or non-cooperative prisoners could be subdued by filling areas with foam 'chemically similar to that used in hair shampoo and liquid soaps' – laced with the agent OC. Among the safety issues are the toxicology of the foaming agent, the OC and the combination of the two acting together, as well as the respiratory implication of being immersed in such foam. Several problem areas were identified in testing. While in general such foam was considered safe, the risk of aspiration pneumonia increased when individuals were intoxicated, using drugs, suffering from seizures and other medical conditions (points that the investigator noted 'may be of concern in dealing with an incarcerated individual'). In addition, the foam would greatly reduce the visibility of individuals caught in it, perhaps resulting in a complete loss of sight. This suggests that officers and inmates would also have trouble communicating with one another, thereby making it difficult to de-escalate the situation or to enable officers to know what effects had taken place. In a situation involving disturbances between multiple inmates, these visual and audio impairments might be major impediments to ensuring an orderly use of the foam. Knocking the foam down quickly in the case of emergencies also proved complicated and required flash-bang grenades or carbon dioxide fire extinguishers. Various decontamination procedures would also be required to clean out the room of OC.

There are, in short, then, a variety of considerations that might come to bear in evaluations. Given the possibility that a small margin might exist between 'acceptable' and 'unacceptable' effects, the basis of determining acceptability is not trivial.

Statements made by government and industry representatives have fostered the perception that many of these non-lethals are designed for progressive purposes and are, in fact, relatively harmless. Initial reasons have already been mentioned for doubting such claims. As mentioned in Chapter 3, for instance, Jürgen Altmann has contended that a number of proposed 'second-generation' weapons are unrealistic.[5] So audible acoustic devices might act to make situations intolerable for individuals, but in doing so they would almost certainly cause long-term hearing loss.

Commentary on the lack of testing regimes has come from many quarters. John Alexander claimed that the lack of comprehensive testing was the 'Achilles Heel' of non-lethal weapons.[6] The Omega Foundation has surveyed regulatory procedures for the approval of non-lethals in Europe and the United States and found no international standards or agreed procedures for determining the safety of non-lethal devices.[7] After noting various unsubstantiated reports about acoustic weapons and their efforts to determine the human effects of such weapons, members of the US Air Force Research Laboratory remarked:

While the research reported here was conducted over a period from 1995 to 2000, it was not initially planned as a six year effort and funding was provided year to year and never certain. Some of the studies were planned over 2–3 years, but other parts of the program were opportunistic based on the availability of acoustic sources. Urgent deadlines, funding limitations, and responding to changes in direction were constant distractions from addressing the scientific issues ... Unfortunately, the approach described here for NLW [non-lethal weapon] acoustic research, is similar to that employed for most NLW bioeffects programs. One reason for this approach is that organizations that have the ability to conduct bioeffects research usually do so to identify hazards, cure diseases, or look for basic mechanisms. There are few laboratories with the mission to systematically and optimally study the incapacitating or aversive qualities of stimuli on humans. For new medical treatments, many years are needed to conceive and elaborate the treatment, test it for safety in animals, and conduct clinical trials before the treatment can be applied. Issues dealing with the effective and safe development of non-lethal weapons are none the less complex, yet they receive much less serious attention.[8]

In the United States, organizations such as the Human Effect Advisory Panel and the Non-Lethal Technology and Academic Research unit have been established with the aim of correcting this overall situation. Much of that work, though, is still in its initial phases. A further complication with many of the latest generation of non-lethals is that they target particular physiology processes and interfere with basic bodily functions in relatively novel ways.

The chapter provides two studies of the health and safety concerns associated with non-lethal weapons. While the next generation of weapons are said to pose the greatest dangers and promises, it is perhaps premature or impractical to discuss many of these. The rest of this chapter considers past and more recent innovations in the major established areas of kinetic and chemical technologies. An evaluation of these conventional, relatively 'low-tech' or 'off-the-shelf' devices provides a basis for asking what procedures might be required for more novel forms of weaponry.

KINETIC-ENERGY WEAPONS

This section considers the safety of impact blunt munitions, designed to deliver a kinetic energy over a distance. As mentioned earlier, the main rationale given for such devices is that the use of traditional firearms by law-enforcement and security agencies is associated with many unfortunate consequences. Munitions and launchers that deliver less force or do so in a less damaging manner are said to offer more appropriate force responses. For instance, former US Army Delta Force soldier and founder of A.L.S. Technologies, David Alvirez, has researched less-lethal impact ammunitions. Promotional material for the company outlines the rationale:

> Alvirez has developed a less-than-lethal bullet product line, which he calls his 'Power Punch.' Delivering a blow similar to what one might get in martial arts, the Power Punch, he said, 'is like being hit with the equivalent of a .32-caliber automatic but it's spread over such a larger area that it doesn't penetrate.' It can knock a person out, he said. 'Part of our statement is that it should never be aimed at the head.' The bullet, for a 12-gauge shotgun, primarily has been used by law enforcement he said, but it was his intention in developing it that it would be for home defense. 'We thought we could provide an alternative to taking a life. If you use this bag, you can say, "I've given it every chance." The next shot is your normal lethal round. This is to take control of the situation so you can allow the authorities to get there. People's attitudes change once they've been hit with that amount of force. It's called attitude adjustment ammunition'.[9]

Police Lieutenant Steve Ijames has commented extensively on the potential of kinetic weapons.[10] He argues that with the proper training and de-escalation techniques, these devices can provide resources for resolving difficult circumstances while reducing injuries and deaths to members of the public and officers. Non-compliant, non-aggressive but armed deranged or suicidal suspects, for instance, can be incapacitated with relatively few dangers. Achieving this, however, requires getting a number of things right: shot placement, energy transferred, projectile design and target mass are each identified as critical factors, though the first is seen as the only one individual officers on the ground are likely to be able to control. In relation to shot placement, areas such as the neck, head and liver are particularly vulnerable, though even the chest can be.

The basic tension as he portrays it is that those areas of the body that are most likely to be injured are also those most likely to incapacitate individuals. With this being so, officers need to selectively target areas of the body (leg versus chest) depending on threat assessments made. The distance between user and target is also crucial; while at long range a given weapon might be safe, firing it up close could be quite dangerous. Thus, a balance needs to be struck – between competing factors that might cause injury and those that achieve incapacitation. Questions about what devices are used, when and how, become important matters. In an effort to provide information for making decisions, Ijames has tested seven versions of projectiles, and supplied information on their properties (such as kinetic-energy impacts).[11]

Despite the availability of forms of non-lethal impact weapons for the last several decades, little agreement has existed about their proper mode of evaluation. After the civil disturbances of the late 1960s, military scientists in the United States established a methodology for assessing the effects. This was primarily in relation to the kinetic energy of weapons – a relatively simple determination based on mass and velocity of projectiles at certain distances.[12] Various bands of impact energy safety were established for the 'average' person of a given weight: levels greater than 122 Joules (90 foot pounds) were in the severe damage range, between 40 and 122 Joules (30 and 90 foot pounds) were deemed 'dangerous', and amounts below 40 Joules (30 foot pounds) were gradually less and less harmful. While a simple basis for evaluation and comparison, these standards would call into question the suitability of many existing options. All seven of the projectiles tested by Ijames, for instance, fall into the severe damage range. In a study of the kinetic energy levels of prominent munitions, the Omega Foundation found that at the target range specified by manufacturers, eleven munitions were in the 'severe damage' range and three were 'dangerous'.[13]

Other techniques have been developed in recent years to assess the likelihood of permanent injuries by including parameters such as terminal momentum, impact duration, contact area, density and shape of munition.[14] Techniques for estimating injuries based on automobile crash testing are being imported into the study of projectiles, and defense companies such as JAYCOR and Defense Technology are developing their own testing systems. Just what methodology is most suitable, though, is a matter of disagreement. A recent research review sponsored by the US National Institute of Justice concluded that while there were numerous techniques for modeling injury, none of them were validated for all projectiles and impact locations.[15]

Harm sustained by kinetic weapons includes minor abrasions and bruising as well as more severe injuries such as broken bones, damage to internal organs, scalping, blinding, fractured skulls and even death. Given the disagreement about the proper models for predicting effects and the possibility that many existing projectiles raise serious safety concerns, information about the field use of such weapons becomes all the more important. In the United States, where individual police forces generally use such technology for a limited range of circumstances, systematic information on the effects of weapons is difficult to obtain. While a number of deaths have been associated with kinetic-energy weapons,[16] some officers have avowed to the relative safety of such devices.[17] Few independent medical studies have been undertaken, though, in part perhaps due to expectations about relatively minor injury levels.[18]

In one such study, a group lead by De Brito studied the effects of the beanbag projectile in southern California.[19] Although such munitions were developed in the early 1970s, they became widely deployed in the mid-1990s. Today, police forces in all fifty American states make some use of this equipment. While various anecdotal accounts of beanbags causing death and serious injury have been made, comprehensive studies have been lacking.[20] The authors documented injuries sustained by individuals brought into the Los Angeles County/University of Southern California hospital between 1996 and early 2000; the likely treatment centers for members of the public from policing encounters with the Los Angeles Sheriff's Department and the Los Angeles Police Department. Both forces require medical examinations after the incident for those hit. There were 40 individual cases, involving 99 beanbags, considered. In over a third of the cases (37.5 per cent), individuals were documented as having suicidal ideation. There were 15 people tested positive for alcohol (though no levels were indicated), 8 for cocaine and 3 for anti-depressants. The injuries included one death, fractures, *penetrating* wounds, facial lacerations and testicular failure, as well as various other serious injuries. In commenting on the findings, the authors said:

> It is clear that the bean bag has not always performed as the design intended … Both the pellets and the entire bag have caused penetrating injuries … Although the possibility of internal organ injury from the bean bag weapon was previously hypothesized, a more impressive spectrum of injury has been demonstrated than was hypothesized. It appears no area of the body is immune from injury.[21]

A 'pattern of underestimation' was seen at work in clinical beanbag injury assessment. The confused or altered mental state of many individuals meant it was difficult to gauge injury through initial diagnoses. The severity of injuries was generally only determined later. As the authors noted, though, the sorts of injury sustained should have been anticipated because as far back as the mid-1970s the US Army had found beanbags had a high probability of undesired effects.[22]

Additional important points were put forward in the study. In a limited number of cases, officers were able and willing to supply firing distances. The doctors speculated that one reason for the high rates of serious injury was that the beanbags were fired under the 9 meter minimum range specified in police guidelines. The use of multiple rounds against individuals, in some cases without effect and in other cases in addition to other non-lethal options, also led the authors to doubt the effectiveness of the beanbag. It was concluded that emergency physicians needed to approach the management of such injuries more carefully than they had in the past.

The findings raise a number of safety-related questions, as follows. Are the injuries incurred regrettable but acceptable given the need for officers to respond to belligerent individuals? Might many of the cases have been handled differently if officers had known the extent of possible injury? Was the firearm the only other option left? What injuries might have been incurred if the beanbag had *not* been available? Could other impact weapons result in less injury? Might beanbags modified for short-range use prove more acceptable? If all cases of the use of this weapon required a medical examination, why had the rate of serious injury not been identified previously by officers using the weapon? How often are multiple shots required and what implications might this entail for safety evaluations based on single rounds? To what extent might the underprediction and clinical underestimation of effects be found with other, so far less rigorously evaluated, kinetic weapons? How many other forces have adopted and set terms of use for the beanbag in light of what this study would suggest are unrealistic expectations about minimal injury? Answers to these questions depend on detailed or counterfactual information that would no doubt provoke much discussion.

Whatever the answers, the Los Angeles study would suggest a rather more complicated set of issues at stake in the evaluation of non-lethal weapons than the question of whether they are safer than conventional firearms. Arguably, the experiment of comparing the penetration depth of a conventional firearm and a bean bag by John Alexander and Sid Heal (Los Angeles Sheriff's Department), mentioned in Chapter 2, is more equivocal in its implications than suggested. The lower penetration depth of the beanbag was seen as 'unambiguously' demonstrating its non-lethal effects as compared to those of a conventional round.

While this relative damage might still hold for *non-penetrating* bean-bags, just what such a test ought to imply in terms of the merits of this technology is open to contrasting evaluations and qualifications.

The inaccuracy of munitions is not a factor mentioned in the Los Angeles study, but might have contributed to the severity of injuries documented. Whatever methodology one uses to predict impact injury, determinations of safety are dependent on shot placement. While individual manufacturers provide accuracy-performance specifications, the comparability of such ratings and their validity has been called into question.[23] Measures of accuracy and the ranges used differ between manufacturers and between types of munition from particular manufacturers. In surveying the range of non-lethal kinetic projectiles, in late 2001, agencies of the UK government contended that manufacturers' data often could not be relied upon to provide correct assessments.[24]

These sentiments are corroborated by an evaluation in 2001 by John Kenny (Pennsylvania State University) and Sid Heal and Mike Grossman (Los Angeles Sheriff's Department) of extended-range impact munitions. Numerous munitions were measured for characteristics such as short- and long-range accuracy, imparted momentum, weight and price. Overall, the tests gave some doubt about the performance characteristics of the projectiles examined. The authors were 'struck by the inaccuracy of these munitions'.[25] At 75 feet, for instance, of the 37 munitions fired, 30 per cent had a dispersal pattern between 18 and 36 inches while 24 per cent could not reliably hit the target plate. Reliability of firing at all proved a problem for some munitions and the imparted momentum varied considerably. Summing up the results, Sid Heal commented:

> The good news is that we have a fairly good idea of what works and what doesn't work, the bad news is that there isn't much that works ... Previously, there had been no such comparison data for these types of munitions, but now for the first time we're able to fire a shot across the bow of the industry ... [26]

Although the evaluation was perhaps the most comprehensive of its kind, a variety of limitations were noted. The authors did not attempt to set standards for what constituted sufficient accuracy or to relate momentum to injury potential; rather, characteristics of different projectiles were measured by common methods and collected for a single document. Furthermore, as only five rounds of each munition were used, the findings were meant to apply to 'munitions as a whole and not about any particular munition type'.[27] Individual police forces making procurement decisions would have to interpret the results for themselves. As John Kenny noted, 'we are not able to predict injury or incapacitation levels to individuals from the results. Our intent is not to make recommendations on certain munitions, but to guide law enforcement in making their own decisions.'[28] Further tests with larger sample sizes were recommended.

Unlike the United States, in other parts of the world kinetic projectiles primarily function as police and military options for public-order incidents, sometimes stemming from internal conflicts. Britain has made some of the longest and most extensive uses of kinetic weapons, first in relation to policing the former colonies but more recently in Northern Ireland. The deployment of first rubber, and then plastic, bullets to the British Army and the Royal Ulster Constabulary (RUC) has been a source of considerable debate, particularly in relation to the hazards of the rounds. Little is known publicly about the official evaluations of rubber or plastic bullets prior to their initial deployment in the 1970s or variations offered since that time. Injuries sustained have included death, permanent disfigurement, fractures, lacerations, eye damage and ruptured internal organs.[29]

Officially, 16 deaths and more than 600 injuries have been attributed to plastic bullets in Northern Ireland.[30] Whether such figures are taken as accurate depends on assumptions made about what constitutes robust evidence. Groups questioning the acceptability of the rounds have suggested that these figures significantly underestimate the level of injury. Requirements on medical personnel to report to the authorities wounds from plastic bullets, and claims about RUC officers patrolling the entrances of casualty units after major confrontations, are said to have led to an underestimation of injury rates. In such settings, wounded individuals were deterred from seeking medical help through public services and turned to make-shift assistance centers.[31]

Whatever the actual injuries, the reasons for why they might have taken place are likewise disputed. As indicated in previous chapters, much of the attention of security forces in Northern Ireland has focused on the properties of the launchers and ammunition. P. A. J. Waddington has referred to the early plastic-bullet gun as 'scandalous' and 'wholly unsatisfactory' because of its unreliability, inaccuracy and difficulty of use.[32] Others have sought explanation elsewhere. Organizations such as the Committee on the Administration of Justice strongly criticized the past lack of controls and the possibility to justify nearly any recourse to plastic bullets, at whatever distance and to whatever part of the body.[33] These questions about where and how to attribute responsibility are given further attention in the next chapter.

As the latest stage in a series of modifications made to plastic-bullet rounds and launchers, in 2001 the Ministry of Defence, the Home Office and the Northern Ireland Office introduced a new round: the L21A1. The Defence Scientific Advisory Council (DSAC) carried out the evaluation by comparing the previous round fired from the L104 gun with the L21A1 fired from the same launcher but with a new optical sight. While details of those tests have been restricted from the public, summary findings have been made available.[34] As concluded by DSAC, with the latest sight, the L21A1 round configuration proved more consistent and accurate, though it had greater risks of causing severe injuries to the brain, of being 'retained' in the head if the round struck perpendicular to the skull, and of ricocheting (thereby hitting non-targeted individuals). DSAC found the risk of life-threatening injuries was reduced, though, because of the improved sighting. Those critical of the new bullet have pointed to the risk for more severe injuries and the unacceptably high kinetic energy of both the previous and the new round (244 joules) at the specified minimum use range of 20 meters.[35] To allay fears of injury, the UK Home Office Minster at the time of the go-ahead insisted that 'the new round, like the old, will be used in situations of public disorder only in accordance with the existing strict guidelines ... [T]he improved accuracy of the new baton round makes it suitable for use in dealing with people who are posing an immediate threat to life in circumstances in which use of a firearm would otherwise be necessary'.[36]

It was assumed throughout DSAC's evaluation that the weapons would be used according to the guidelines. Commenting on the possible effects of the new system in practice in relation to the guidelines for use, DSAC noted:

> that it may be difficult to maintain the acceptable incidence of injury at the low level currently envisaged, in all *operational* as distinct from *test and training* circumstances. We emphasise that the ... recommendations are critically predicted on such assumptions of acceptable competent training: this needs to be kept in mind by those who make policy decisions.

Given such qualifications, the DSAC highlighted the need for further medical studies and the monitoring of the use of the L21A1.

This review of recent debates about kinetic-energy impact projectiles suggests a number of points. Despite the widespread deployment of this class of non-lethals for a number of years, predictive evaluations and assessment of operational effects have been limited. The conclusions of many evaluations are highly qualified. Ensuring that the kinetic-impact weapons function as advertised requires that various assumptions about their usage be held. Moreover, the acceptability of such devices is gauged in relation to considerations about the threat faced and alternative responses. When things are identified as having gone wrong, identifying the source cause of this (for instance training, faulty technology, procurement decisions or guidelines) is a matter of potential disagreement. Implications of this section are taken up later in this chapter after a consideration of another non-lethal technology.

CHEMICAL INCAPACITANT SPRAYS

Chemical agents such as CN, CS, CR and OC, mentioned in Chapter 3, have been incorporated into aerosol canisters, sprays and other devices. Supportive claims about the class of technology suggest that the agents' irritating and disabling effects are temporary and recede shortly after contact stops. Yet since the introduction of agents into military and policing operations early in the twentieth century, questions have been asked about the safety of weapons using such chemicals. As with the kinetic-impact projectiles mentioned above, an examination of the claims about such chemical non-lethals raises a host of issues about how determinations of safety are made.

A broad review of the history of chemical agents indicates something of the scope for negotiation. Take, for instance, the case of CN. It was one of the first chemical agents to be incorporated into 'tear gas' devices. Tests conducted at the UK defence laboratory, Porton Down, in the 1920s initially suggested that there were few lasting effects from exposure to this chemical. Complaints made at the time by those exposed about lasting complications (such as dermatitis) were dismissed as inaccurate or aberrations. On the basis of thousands of testing exposures, CN was considered to cause no lasting injuries in Britain and the United States. Gradually, that assessment began to change. By the 1950s, CN had been linked to serious skin and eye injuries, though its use as a riot-control agent in the British colonies continued until the mid-1960s.[37] CN is now considered an unacceptable option in Britain. Much the same story of shifting evaluations could be told for the case of CR.[38] Key parameters about the safety of chemical irritants relate to the dosages received and the methods of its dissemination.

Establishing the link between exposures and long-term medical conditions is difficult and resource-intensive. In 1971, a doctor working for the Washington, DC Police and Firemen's Clinic found that 12 patrol officers had developed a rare form of skin cancer known as malignant melanoma.[39] These officers had been exposed to tear gas during the policing of protests during the period 1968–71. Believing that the gas caused no lasting damage, some of the officers had not used gas masks because it impaired their vision and hearing. Allegations were made at the time of widespread and excessive use of such chemical non-lethals in protests. So much was used that the police force ran low in their stock of recent tear-gas grenades and resorted to previously discarded grenades from the 1930s. The police doctor attributed the cancer to the older versions of the tear-gas devices (almost certainly CN), though tests to establish this were never conducted due to a lack of research facilities.[40]

Concerns have been voiced about the potential for variable and unknown effects due to the complexity of chemical interactions. CS was first used in the United Kingdom on a large scale in Northern Ireland. After the 1969 Derry riots, the Himsworth Committee of inquiry was set up to look at the medical and toxicological effects of CS smoke (or 'gas'). While generally supportive of the relative safety of CS against healthy persons in open spaces, the Committee acknowledged significant limitations in knowledge. Noting the uncertainties and the possibility for varied effects across the population, it suggested:

> if the competent authorities feel it justifiable to release a chemical agent for use in civil circumstances, the medical and scientific research relevant to this decision should straight away be published in the appropriate scientific journals so that informed medical and scientific opinion may assess the situation for itself.[41]

This suggestion was most likely prompted in the case of Northern Ireland because experiments to assess the carcinogenicity of CS were completed two years after it was extensively employed there.[42] The Committee, as with suggestions made above about acoustic weapons, further recommended chemical agents like CS should be regarded as more akin to drugs than weapons for the purposes of their approval. The findings of the Committee have served as the official benchmark in Britain for evaluating riot-control chemicals. As we will see in later chapters, alternative interpretations have been offered of just what was meant by the suggestion of regulating 'akin' to a drug (see Chapter 8).

While aerosol or gas riot-control agents were deployed throughout the twentieth century for crowd control, it was only in the last thirty years of the century that personal chemical incapacitant sprays became readily available in some countries, and only in the last few years that their adoption became widespread. Chemical sprays typically consist of three components, the primary chemical agent, a liquid solvent and a propellant gas. Although in some countries such sprays are available to civilians (in less concentrated forms), by and large the main users are law-enforcement agencies. Such sprays are said to act as an effective option in handling unruly individuals from a distance, a deterrent in assaults against officers and a factor in the reduction of excessive-force complaints against the police.[43]

Police and correctional officers in the United States make some of the most extensive use of chemical sprays, primarily OC-based sprays (oleoresin capsicum – or 'pepper spray'). While such sprays have been available since the 1970s, their widespread use did not begin until the early 1990s. The principal active agent of OC sprays is capsaicin, an extract of peppers. Those sprayed generally experience an intensive burning and swelling in their airways and eyes that makes physical exertion difficult. Eyelids are forced shut, blood rushes to the head and the mucous membranes secrete fluids. Because of the temporality of these effects, pepper sprays are said to represent a humane alternative to firearms or nightsticks.

During the recent surge in the uptake of pepper sprays, there have been fairly persistent concerns regarding the lack of knowledge about their health effects, the possibility for overexposure, and the variability of effects due to pre-existing medical conditions. Among the most prominent, in 1993 members of the US Army Edgewood Center concluded that limited safety tests had been conducted on capsaicin and what information was available suggested its use on a varied population posed significant risks. Negative implications included sensitization, mutagenic and carcinogenic effects as well as cardiovascular and pulmonary toxicity.[44] The extent to which such risks would lead to health problems would only become apparent after many years of usage and exposure. Another major criticism has come from the American Civil Liberties Union and others who have 'associated' a number of police in-custody deaths with pepper sprays.[45]

Part of the difficulty with assessing the health effects of OC is that there is a variety of types of such devices currently available. Spray brands differ in terms of their concentration of capsaicin, the concentration of this extract in sprays (varying from 1 to 10 percent), the types of solvent used (water, alcohols, organic solvents), the particular methods of dispersion (stream, spray or fogger systems). OC itself, being derived from a natural product, is a mixture of numerous volatile compounds. This status as 'natural' has also meant that OC has not been subject to significant national regulatory controls in the United States (though certain states adopt limited testing requirements), and to considerably fewer controls than are applicable to human-made chemicals such as CS.

Although the sprays are supposed to be safe when applied properly, when things go wrong there are questions about where blame rests. The variations in pepper spray suppliers, in rules of engagement and treatment and in police practices bring ample scope for disagreement over the attribution of blame. One of the most famous such instances of this took place in 1993. A pathologist recorded that a man in North Carolina died from asphyxia due to bronchospasm precipitated by pepper spray. For a time, news of the event led to the withdrawal of pepper sprays from a number of forces. According to police trainer, Mike Doubet, the spray manufacturer in question laid down the cause of death to the failure of the affected individual to receive the necessary medical treatment. In other words, the fault was with the conduct of officers and not the sprays 'themselves'. Noting the scope for distributing responsibility and liability, Doubet concluded:

> These comments by a manufacturer blaming the police for the death and clearing their product of any responsibility should be more than alarming to agencies and officers alike. With no medical or scientific proof, they make a public statement laying blame on the officers based on what their investigators concluded. With a lack of research on their products and less than unbiased investigators looking into cases for manufacturers, [police] agencies should begin to realize how much support they will be given in court by the manufacturers should a lawsuit arise.[46]

Despite such warnings, others have made highly optimistic claims about the potential of OC sprays. US National Institute of Justice-funded researchers, Smith and Alpert, found that pepper spray 'seemed to be an ideal solution to the problems of excessive force because it effectively controls resistive suspects while inflicting little or no lasting injury'.[47] In the last decade, OC sprays have been taken up in countries such as Australia, Belgium, New Zealand, Finland, Canada and Russia.[48]

In 2000, the Netherlands became one of the latest countries to deploy OC sprays. In 1998, and in contrast to practices in many other countries, researchers sponsored by the Ministry of Internal Affairs published an article detailing the risk assessment of the sprays.[49] This provides a convenient case for examining how claims and counterclaims about the safety of the sprays were handled. Those researchers concluded that while some risks were associated with OC spray, as a police weapon it 'is safe, meaning that brief exposures to relatively low amounts of OC, as expected in this context, does not result in irreversible damage to airways, skin, or eyes'.[50] Evidential support for this contention was drawn from a number of sources. It was said that though little was known about the acute toxicity of OC inhaled by animals, past experience with humans suggested that concerns here were limited. Studies by the Federal Bureau of Investigation (FBI) Firearms Training Unit in the late 1980s, with around 1000 volunteers, for instance, indicated almost no concerns for side effects or adverse reactions. A much smaller study of 81 people found that OC exposure resulted in no hospitalization.[51] The potential for deaths in police custody was downplayed largely through a study conducted by Granfield, Onnen and Petty.[52] The Dutch researchers also surveyed various articles about the mutagenicity and carcinogenity of capsaici-noids, which were said to provide conflicting appraisals. Because of the proposed limited exposure to the sprays, these potential dangers were seen as only theoretical. The risks for people with chronic lung disease were presented as a possible area of concern and one in need of further research. Against the health risks of OC sprays, though, the authors contended that the sprays' effectiveness in subduing suspects had to be considered. Various summary accounts of the operational use of the sprays were cited, especially a study of the Baltimore County Police Department that found the devices effective in 90 per cent of the cases.

A later study published in 2001 for the Dutch Ministry of Internal Affairs followed some points raised in 1998.[53] Despite the sprays being deployed in a number of countries, the investigators found that little research existed about the safety of the sprays for asthmatics and other at-risk populations. Various experiments were conducted with asthmatic and non-asthmatic guinea pigs to determine their sensitivity. The tentative conclusion was made that pepper spray 'exposure induces at most mild bronchoconstriction', provided that the devices were designed to make only a low fraction of the spray inhalable (in relation to particle size), that exposure time was only a matter of seconds (so individuals affected should be moved immediately from the exposure site), and that the sprays were fired at a distance of two meters. The results were deemed tentative because of the difficulty of extrapolating from animal studies the effects for humans. Recommendations were made for experiments on asthmatic human volunteers and funding provisions made, though no such tests have been conducted due to the inability to gain approval from the necessary ethical committee.[54]

As described above, several key studies underpinned the overall positive assessment made of OC sprays in the Dutch risk assessment: the FBI Firearms Unit tests on human volunteers which found almost no indication of adverse side effects, the study by Granfield, Onnen and Petty refuting a casual link between the sprays and in-custody deaths, and operational effectiveness studies of pepper sprays conducted in Baltimore. The remainder of this section examines these studies in more detail.

A Rereading of the Canonical Texts

FBI Firearms Training Unit: Research undertaken during the period 1987–89 by the FBI provided one of the first systematic studies of OC sprays. Evaluations were conducted with a version of the spray called CAP-STUN, then manufactured by Luckey Police Products. In 1996, however, the agent in charge of the FBI tests pleaded guilty to a felony charge due to his wife's firm receiving $57,000 from Luckey Police Products in 1990.[55]

Until 1996, manufacturers and commentators cited these studies as evidence of the safety of OC sprays.[56] The FBI studies provided justification against critical assessments offered, such as those by the US Army Edgewood Chemical Research and Development Center. In 1991, the company, Zarc International, bought all the intellectual property rights to CAP-STUN from Luckey Police Products. Zarc International was not involved in the bribe in any way. 1993 promotional material for the company described the FBI studies in the following terms:

> Although CAP-STUN® was used by law enforcement agencies to a limited degree in the early [1980s], it was not until the summer of 1989 that use of CAP-STUN® skyrocketed because of the FBI search for a legitimate and effective alternative to use of excessive force and tear gas. In 1989, the Firearms Training Unit (FTU) of the FBI Academy (Quantico, VA), with the assistance of the US Army Chemical Research and Development Center and with the technical assistance of ZARC® completed its three-year study on CAP-STUN® brand aerosol weapons. As a result of its research, the FBI began to issue CAP-STUN® to its agents and SWAT teams. The FBI report on the study, 'Oleoresin Capsicum Training and Use', was widely distributed among law enforcement agencies and served as a catalyst for CAP-STUN®'s growth.[57]

The court-ruling cast doubt about the validity of the FTU findings. A follow-up review of the FTU study by the FBI after the bribe charges found 'that there is no reason to believe that the FTU study did not accurately report the observations made of the effects which the OC exposure had on those tested'.[58] Limitations of the FTU study were acknowledged, such as the lack of testing on those with pre-existing health problems. Yet anecdotal experiences with OC sprays were taken by the FBI to indicate that the sprays were safe.

Alternative assessments have been offered of the implications of the bribery case. For some, the incident raised fundamental doubts about the rigorousness of the evaluation of OC sprays and their likely safety given other critical evaluations made.[59] At a minimum, the controversy over the tests would suggest that their unqualified citation would be unadvisable. Yet, as in the Dutch risk assessment, bald findings of the FTU tests have been cited as evidence for OC spray's safety,[60] even in a technical report for the National Institute of Justice.[61]

In-Custody Deaths: One of the most controversial aspects of pepper sprays has been their contribution to deaths in police custody. Newspaper and other accounts have associated a number of deaths with pepper sprays;[62] 'association' in such statements has generally referred to an unspecified link between the incapacitant and death, something between coincidence and direct causation. Just what link might exist has been a matter of some dispute. In confrontations resulting in death, several force options might be employed by officers, such as batons, physical combat, and restraint techniques.

In 1994, Granfield, Onnen and Petty reviewed a number of cases involving police in-custody deaths and pepper sprays for the US-based International Association of Chiefs of Police and the National Institute of Justice. The authors examined 22 deaths during the early 1990s. A variety of factors that might complicate determinations of the cause of death were noted, such as difficulty of interpreting pathological evidence, as well as the potential of such factors to be used to inappropriately attribute blame or incompetence to the police or medical examiners. Granfield and colleagues drew attention to the importance of contextual features, such as the sequence of the confrontation and circumstances surrounding the deaths, which needed to be brought to bear in interpreting evidence.

Despite the difficulties noted in determining death, several definitive conclusions were given. Positional asphyxia was deemed to be the cause of death in 18 instances, with cocaine abuse, cocaine-induced excited delirium, and neuroleptic malignant syndrome also being cited. The sprays were said not to contribute to any of the deaths studied in any way. The reasoning behind these conclusions was not given, though. No evidence is cited as to why the sprays could not have contributed to these deaths, such as by affecting breathing or by making individuals more excited, both of which would exacerbate responses related to positional asphyxia. Some reasoning is given for this elsewhere. Doubet quotes an article interview where Granfield is reported to have said: 'Regardless of what method was used to subdue the subject, whether it's the night stick, a choke hold, or five officers sitting on the guy, none of these things really matter, it was more the individual's profile that was going to determine whether or not the subject might die suddenly in police custody.'[63] In one case where the OC spray was considered as a factor in death in the autopsy report, Granfield, Onnen and Petty refuted this contribution, though no grounds were given.

The lack of evidential argumentation, though, has not stopped the results of this review from being cited without qualification in a number of evaluations of pepper spray.[64] The evidence for determinations is critical, though, as the understanding of deaths changes over time. In the Granfield, Onnen and Petty study, five of the people who died were restrained through 'hog-tying' techniques (where individuals lie on the stomach with their hands and ankles tied behind their back). It is not specified whether any of those individuals 'hog-tied' had their deaths attributed to positional asphyxia. The potential for positional asphyxia from such a restraint technique was downplayed in 1998 by a study by Chan, Vilke, Neuman and Clausen from the University of California San Diego Medical Center.[65] While this study has served as an important piece of evidence for legal proceedings, it also cast doubt about the validity of past determinations of death by positional asphyxia.

With further regard to evidential requirements, others have taken a more cautious view than Granfield, Onnen and Petty about the potential for pepper sprays to contribute to deaths. A study by Steffee, Lantz, Flannagan, Thompson and Jason[66] presented the details of two in-custody deaths, suggesting a direct contribution of pepper spray to death. The authors drew attention to the specific information required for determining the cause of in-custody deaths. These include the circumstances of arrest, the deployment of OC sprays (number of times used, particular spray's characteristics), physical confrontations and injuries before and after the event, evidence of intoxication, symptoms of illness before and after exposure, the history of individuals' social, mental and physical well-being, the position during transport, autopsy findings, toxicological results, as well as attention to the surveillance procedures in place for collecting such information. Determinations of death caused by or contributed to by pepper spray were seen as requiring a careful weighing of such issues, where (as in the two case studies discussed) the evidence needed to be presented and debated 'to reveal the role of OC spray as unrelated, contributory, or causative'.[67]

In December 2001, the role of OC spray as contributory or causative was challenged by a National Institute of Justice study conducted by Chan, Vilke, Clausen, Clark, Schmidt, Snowden and Neuman. These researchers examined the effects on 34 officers of being sprayed with OC, in both a sitting and a restrained position. They found little evidence that OC exposure caused a 'significant risk for respiratory compromise or asphyxiation, even combined with positional restraint'.[68] Several limitations of the test were noted by the authors: the tests were done on healthy police volunteers not under the effect of illicit drugs; none were sprayed in the eyes, thus possibly aggravating his or her response; the sprays were administered according to manufacturers' guidelines – one spray burst from five feet away; exposure time to the area being sprayed was restricted to five seconds; the limited number of overweight individuals or those with respiratory abnormalities meant definite findings could not be reached for these groups; and none of the individuals were agitated, excited or exhausted from drugs or physical exertion as might be the case in confrontations with the police. On the negative side, some evidence was found that OC exposure did result in increased blood pressure; a result whose implications were said to be unknown. Despite the limitations, it was concluded that the study provided law-enforcement agencies with much needed clinically-based evidence of the safety of OC sprays.

There would seem to be some grounds for asking whether, because of these limitations, the results of the study have much relevance for practical uses of OC sprays. For instance, in the Granfield, Onnen and Petty study, three-quarters of the cases studied involved alcohol or drugs.[69] Added to the limitations acknowledged by the authors are the likely prior familiarity of at least some officers to pepper spray and the examination of only one form of restraint: the 'hog-tied' position. In addition, as mentioned above, some of the same medical authors that conducted the NIJ study had concluded earlier that hog-tying did not result in significant ventilation changes. Interpreting the significance of this study requires thinking through the implications of experiments in an arguably artificial setting, done in accordance with guidelines, to individuals in different states of health than those identified to be most at risk, and for a form of restraint itself not associated with significant ventilation implications.

Effectiveness Studies: Another pillar of support for OC sprays cited in the Dutch evaluation was a National Institute of Justice-supported study undertaken by Edwards, Granfield and Onnen of the sprays' use in the Baltimore Police Department.[70] The study was based on results obtained from use-of-force data sheets and interviews with officers between July 1993 and March 1994. The authors made various highly supportive claims, including that the OC sprays proved effective in 90 per cent of cases.

The initial work in Baltimore was followed up in 1999 by a larger study by Kaminski, Edwards and Johnson.[71] The second drew on 700 incident reports involving pepper spray between April 1994 and December 1996, in contrast to the 174 that served as the basis for the initial study. The findings of the latter one gave a less optimistic appraisal. Effectiveness was assessed by three officer-based standards: whether the spray eased arrest, whether it incapacitated suspects, and what the behavior of suspects was after being sprayed. The authors concluded that 'researchers can obtain substantively different results regarding the effects of explanatory factors when using different criteria ...'.[72] Asking whether suspects were incapacitated or not gave a 70.7 percent incapacitant rate. In contrast to other use-of-force options, this is relatively low and certainly lower than previous determinations that fell in the range of 90–100 percent. Using the criterion of easing arrest, the sprays were considered effective 85.3 percent of the time.

The scope for alternative effectiveness criteria to lead to different assessments of effectiveness is illustrated in a 1997 review conducted by the Berkeley Police Review Commission.[73] Based on its analysis of 36 encounters, they found OC spray to be ineffective 53 percent of the time and fully effective only 20 percent of the time. In making its determinations about effectiveness, the Commission labeled cases where subjects continued to struggle as 'ineffective'.

Considering possible reasons for ineffectiveness, retired officer Phil Messina has argued that the effects of the sprays depend on the mind-set of those targeted.[74] Typically, manufacturers' statements are based on tests with passive volunteers in static 'confrontations'. For a violent or determined attacker, Messina contends that OC and other chemical sprays are almost completely ineffective. In other words, the sprays are likely to fail when they are needed the most. As such, users must be made aware of the limitations of this option. These cautionary points suggest something of the importance of the standards employed for evaluating evidence that predicates statements about the effectiveness of the sprays.

In re-examining the main texts that served as the basis for the Dutch assessment, it is possible to see something of the unstated assumptions, qualifications and contingencies that can underlie determinations. Assessments of what is safe and effective are defined in relation to one another. So, while that noting that various aspects of effectiveness were lower than expected, Kaminski, Edwards and Johnson concluded that OC spray 'remains an important tool for law enforcement, provided health concerns are not a major issue'.[75] Those evaluating health concerns have made parallel conditional statements. Smith and Stopford argued that 'there is no real scientific basis for the claim that OC sprays are relatively safe. In fact, a number of reports have associated serious adverse sequelae, including death, with legitimate use, as well as misuse and abuse, of these sprays.'[76] While suggesting various eye, skin and respiratory risks, on the basis of field studies such as that in Baltimore, Smith and Stopford concluded that 'properly used OC can be effective and provide additional safety to officers'. The division of expertise between medical and operational studies means key issues about the overall evaluation of pepper spray are unaddressed.

Much of the debate about the risks of OC sprays turns on questions about the specifics of exposure. Proponents of the sprays have called the relevance of the Edgewood Chemical Research and Development Center's critical evaluation of capsaicin into doubt. In a 1996 letter to the American Civil Liberties Union, a US Department of Justice official reported that the FBI had determined the Edgewood study to rely on 'high and multiple dosage protocols and failed to address the realistic toxic risk from the much lower doses and exposures to which individuals would be subject in a field setting'.[77] Implicit or unarticulated assumptions about what counts as realistic exposure levels underpin many safety evaluations. The sprays might be deemed relatively safe when exposure is from a single shot sprayed from the required distance and with short-term exposure. Yet, what happens where multiple exposures take place? What happens when a juvenile is sprayed 15 times within a few months?[78] Does this fall in the range of a realistic exposure level?

In relation to at-risk populations, Smith and Stopford argue that due to multiple exposures from training and field usage, there are particular concerns for police officers.[79] In order to familiarize officers with the effects of the incapacitant devices, some US police departments require them to be sprayed directly in the face. On the basis of an investigation of training procedures, in 1996, North Carolina Department of Health and Human Services and the Department of Labor concluded that training exposure to OC sprays entailed an 'unacceptable health risk'. Symptoms included chest pain, headache, asthma attacks and eye irritation, some of which lasted for over a week. Other states have followed suit with various precautionary measures such as the screening of officers for medical conditions and alternative training procedures.

The extent of exposure to both members of the public and police, in part, depends on official operational guidelines. These vary between forces. The sprays might be approved after the failure of verbal commands, physical tactics (grabbing, pressure point control), or other weapons such as the baton. Like the case of the beanbag earlier, manufacturers and others note a greater risk for injury in close-proximity confrontations. In Baltimore, for instance, the sprays were supposed to be used at a distance of over three feet, yet officers' accounts in 60 percent of the time recorded spraying under 3 feet.[80] Such estimations of usage call into question the assumption guiding many safety assessments.

It seems reasonable to assume that readiness to use is also dependent on perceptions of safety. In Baltimore, officers were told that 'In over 15 years of field experience, there has not been a substantiated instance of adverse reaction to the [OC] sprays by any subject.'[81] Others would disagree. In 1994, members of the UK Chemical & Biological Defence Establishment at Porton Down conducted a review of possible riot-control agents for incorporation into sprays for British police forces. At the time, they found data on OC's safety as lacking. Serious concerns were expressed about police-custody deaths. CS was deemed much more appropriate for sprays, though in the United States, OC replaced CS in many forces in the 1990s. A reluctance to approve OC continued in Britain in spite of questions about 'CS' (see Chapter 8).

Much of the previous debate has been pitched in terms of OC versus other chemicals. To speak of 'OC spray', though, as one type of device is to ignore the variations within this category. Smith and Stopford contend that sprays differ 30-fold in risks because of variations in the capsaicinoid concentration of extracts used and the concentration of extracts.[82] Elsewhere, Stopford has been quoted as saying that much of the problem with OC sprays derived from liquid-stream dispersal mechanisms that meant exposure levels were quite high.[83] The lack of oversight in the United States, though, means that such variations are not regulated in terms of their safety. It is not difficult to find OC spray manufacturers who criticize the safety of such devices in their marketing literature, just not their own. Promotional material for Zarc International, for instance, notes the potential for injury and blindness in sprays that mix OC and CS, as well as the dangers associated with solvents in competitors' sprays. Defense Technology/Federal Laboratories conducted their own reviews, but drawing different conclusions than those of other manufacturers about the safety.[84] Recently, many European police agencies have adopted PAVA (pelargonic acid vanillylamide), a type of spray related to OC. Unlike OC, though, PAVA is synthetic and therefore does not consist of a mixture of chemicals.

As with the kinetic-energy weapons in the last section, examining something of the disputes about chemical non-lethals suggests a range of pertinent issues. The safety of these sprays is contingent on various assumptions about the manner of their use. Alternative standards and criteria for evaluation offer contrasting ways of understanding the merits of this technology. The basis of what counts as robust evidence is central to assessments. To the extent that there is praise or blame to be cast, that may be seen to fall with police forces employing weapons, manufacturers designing and promoting them, the characteristics of individuals or absentee regulators.

LIMITATIONS OF ANALYSIS

The previous sections surveyed something of the debate about the safety of two well-established types of non-lethal weapons. As opposed to the highly speculative assertions about future weaponry, it might be expected that experience with 'off-the-shelf' devices would provide a fairly clear-cut assessment. Yet, even taking technology that has been around for some time, continuing questions about their safety can and are being made.

This section elaborates some of the limitations associated with analyses – such as medical predictive evaluations or technological assessments – to answer questions about the safety and hazards of non-lethal weapons. I want to do so at three levels. At the first and most basic level, little is known about the health and safety risks of many non-lethal weapons. That might be because the information does not exist, or it is circulated to a narrow clientele. Governments such as those of the United States and Britain have shown a reluctance to impose regulatory controls that might compel the production and release of testing information. Doubet argues that this environment has enabled pepper-spray manufacturers in the United States to make unsubstantiated and erroneous safety claims about their products.[85] The phrase 'Show Me the Data' has been made at international non-lethals conferences for those such as Michael Murphy and Jürgen Altmann who are doubtful of highly optimistic claims about the possibility of the next generation of non-lethals.[86] As suggested here, though, this phrase could just as well apply to past technologies.

The importance of missing data extends beyond attempts to debunk particular claims as specious or infeasible. As suggested in relation to kinetic weapons, for instance, basic but systematic tests are only starting and operational monitoring practices are limited. Even taking the need for some sort of kinetic-projectile weapon as a given, important questions can be asked about which type of projectile ought to be used, under what situations, and with what provisos. These points are crucial for security forces debating the merits of alternative types of weapon and what guidelines for their use should be given.

The vacuum of expertise in which many deliberations take place may be partially filled in the coming years as current evaluation programs in the United States and Europe come to fruition. It appears that many of the latest-generation non-lethals will be subject to more stringent controls than their predecessors. Yet, even with official proclamations, it is unlikely that debates will simply fade away. In considering some of the limits of analysis, at a second level, the validity, conditionality and scope of claims are key matters. Even if reasonable aggregated proba-

1. First Generation Chemical Irritants (courtesy of Robin Ballantyne, Omega Foundation).

2. Maintaining Order: Tiananmen Square Guard with Electroshock Device (courtesy of Robin Ballantyne, Omega Foundation).

3. An Enlightened Weapon?: The Dazzling Laser (courtesy of Robin Ballantyne, Omega Foundation).

4. A Sign for the Times?: US Joint Non-Lethal Weapons Directorate Headquarters (Quantico, VA) (courtesy of Robin Ballantyne, Omega Foundation).

5. Northern Ireland Baton Rounds: L5A7 and L21A1 (right),
   (courtesy of Robin Ballantyne, Omega Foundation).

6. Mural for Julie Livingstone (Belfast, Northern Ireland), (courtesy
   of Robin Ballantyne, Omega Foundation).

7. Automation for the People: Robot Armed With Non-lethal Gun
   and Lethal Rocket Launcher, (courtesy of Robin Ballantyne,
   Omega Foundation).

8. Non-Lethal Weapon Launchers: Projectiles Available Include Malodorants that Smell of Cadavers, (courtesy of Robin Ballantyne, Omega Foundation).

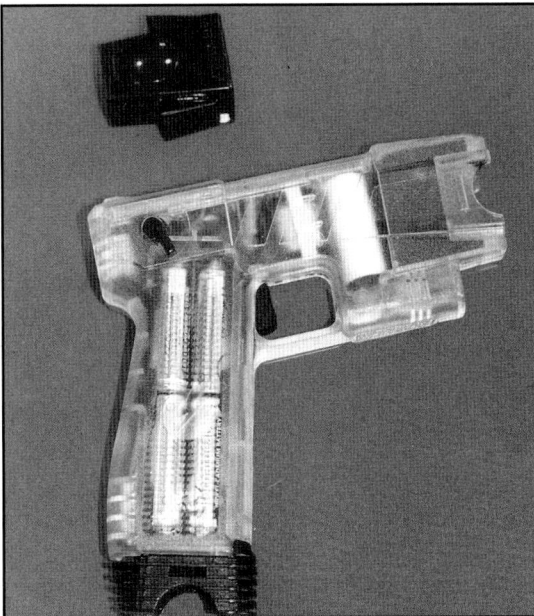

9. The Electroshock Taser: A Transparent Technology? (courtesy of Robin Ballantyne, Omega Foundation).

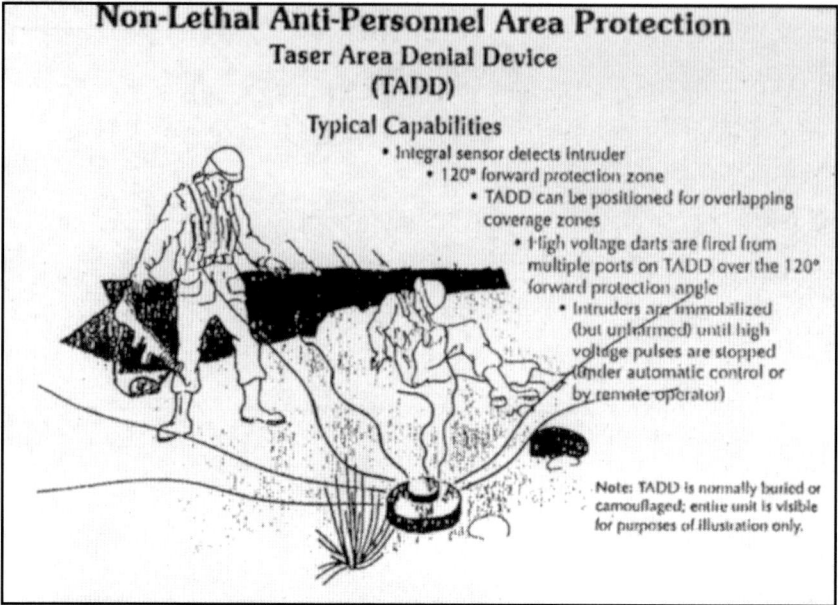

10. The Future of Landmines? (courtesy of Defense Technical Information Center (DTIC)).

11. A View from the Receiving End? A Glossy Slide Presentation about a US Marines Prototype Microwave Directed Energy Weapon (courtesy of Defense Technical Information Center (DTIC)).

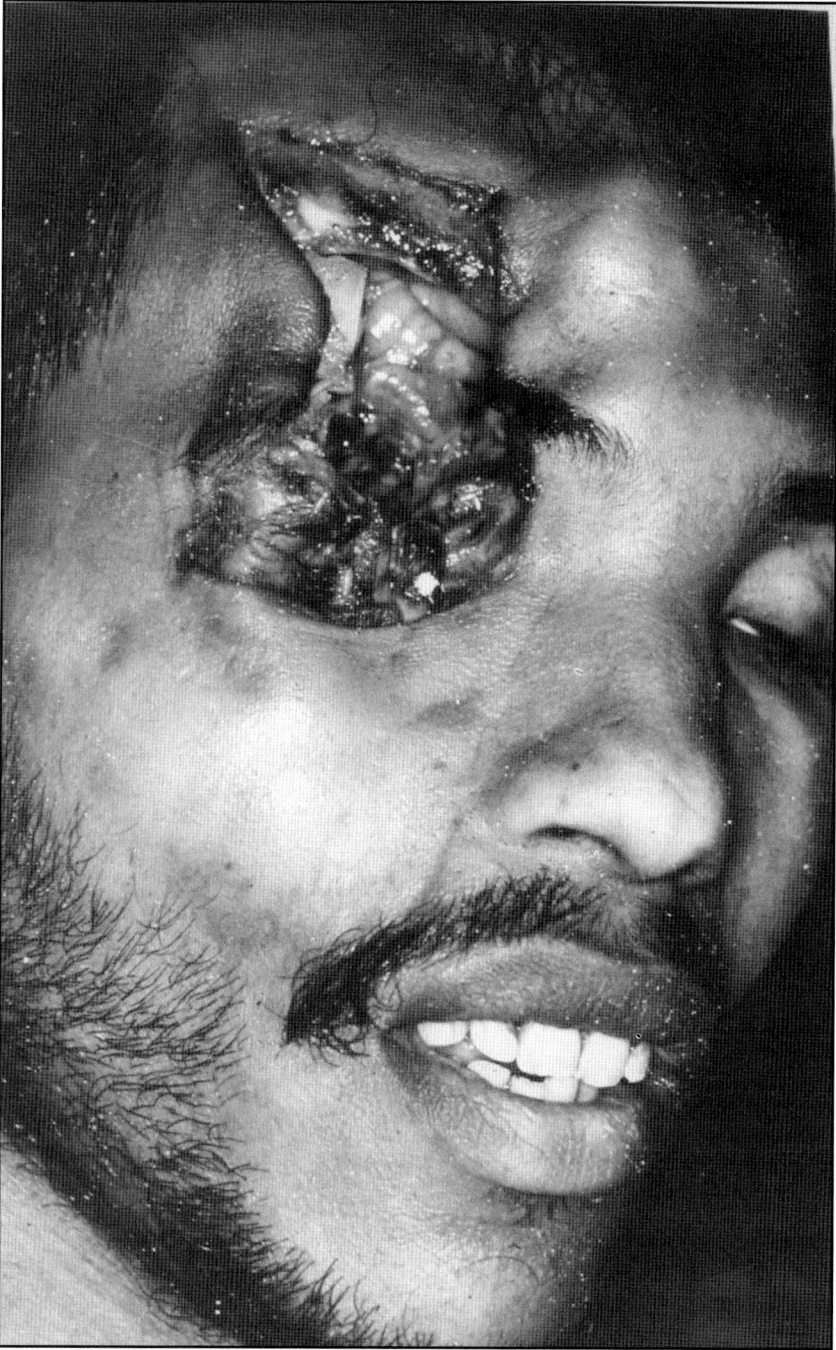

12. Palestinian Rubber Bullet Casualty (courtesy of Tzipi Kahana).

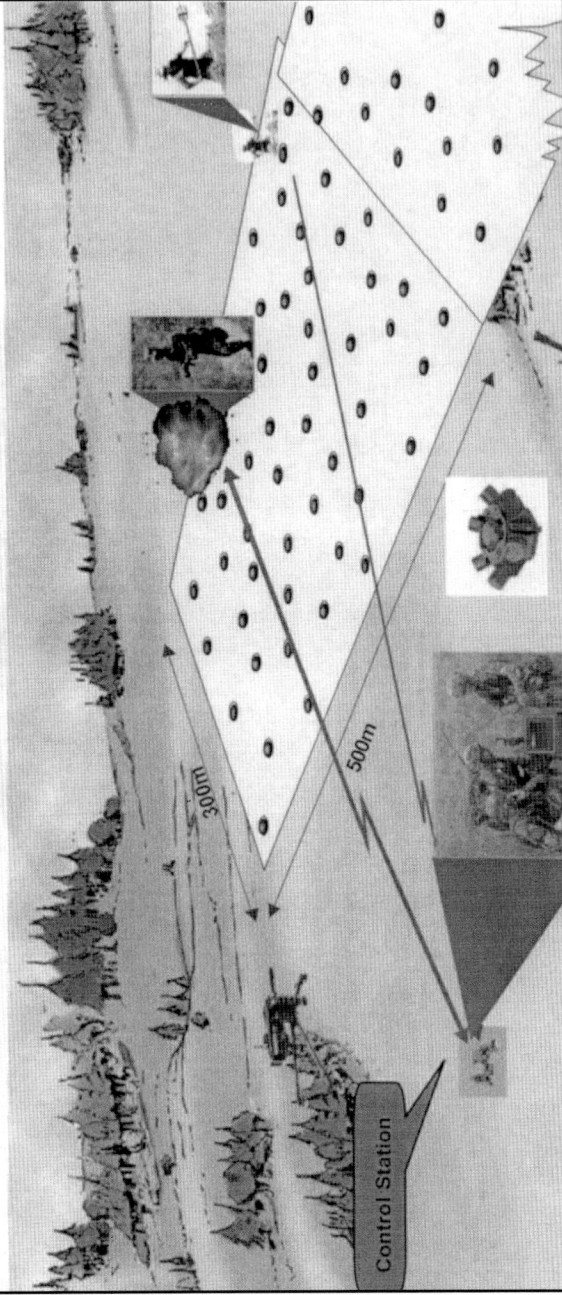

AD-P Operational Concept "System-of-Systems"

Capable of Transitioning from Non-lethal to Lethal
as The Situation Dictates

Control Station

300m

500m

13. A Flexible Response: Not Less/Or but And/More – US Army Area Denial to Personnel (AD-P) Vision for the Future of Landmines. (Courtesy of Defense Technical Information Center (DTIC).)

bilities of injury or death could be predicted or obtained from field experience, such numbers would mean little divorced from contextual qualifications. As argued in Chapter 2, abstracted aggregated figures on their own are not very helpful in assessing benefits and risks. In the case of incapacitant sprays, highly enthusiastic and unqualified claims about the overall effectiveness of the devices (by, say, failing to note their limitations against aggressive individuals) can have tragic consequences for those reliant on the accuracy of such claims.[87] In a similar vein, action has been taken in the United States against at least one company to stop it from making unsubstantiated representations about the general effectiveness of its public self-protection sprays.[88]

Many safety statements are predicted upon assumptions about the manner of a weapon's use and who is targeted. Various 'ifs' and 'buts' accompany promotional literature. The conditionality of the safety of non-lethals and the possibility that a small margin exists between acceptable and unacceptable effects means these technologies are likely to be at the center of various disputes. In any situation where unacceptable injuries arise, questions will need to be asked about the source of these outcomes, whether that be the characteristics of the weapon, the user, the recipient, the general situation, etc. While determinations might be made, there seems ample scope for disagreement. As suggested above, once one examines the assessments made of non-lethal weapons in detail, numerous uncertainties and qualifications can be suggested. Asking one question can spawn many more.

Arguably, many of the problems identified for assessments apply more widely than to non-lethal weapons. Political scientists Lindblom and Woodhouse have argued that analysis in public policy suffers from many limitations: it is fallible and generally recognized as such, it cannot resolve questions of values, it is slow and costly, and it cannot establish what problems to address.[89] This being so, the consequences from assessments are rarely straightforward. For Woodhouse, the assessment of technology is particularly difficult because of the initial unpredictability of how both the technology and its operating environment will function, and because of the long time-lag that often exists between a new technological initiative and detecting its undesirable effects.[90]

In this regard, for a number of years the kinetic-energy-level studies conducted by the US Army in the 1970s were seen as providing approximate, though imperfect, measures of the lethality of projectiles. The renewed interest in non-lethals in recent years has brought greater attention to establishing dangers associated with kinetic weapons. That has not, though, brought consensus on proper methodologies for assessment. Rather than using the established measures of lethality to characterize some weapons as unacceptable (as many of them would be), the practice has been one of waiting for the results of some future consensus to provide a benchmark – though manufacturers whose projects fall within the Army's safe range have taken up this methodology.[91]

Even if agreement could be established about the risks associated with specific weapons in particular circumstances, what constitutes an acceptable level of risk is not resolvable through expert analysis alone. At least in policing situations, individuals who may be the targets have rights that mean they should be protected from unreasonable harm rather than just being dealt with in the most expedient fashion. Yet, in the case of the non-lethal weapons discussed earlier, the risks entailed are not agreed upon. As in many disputes about technology and policy, experts disagree. Despite the major shift to OC sprays in the United States in the 1990s, the agencies of the UK government remain committed to CS. Disagreement also exists about the basic methodologies for assessing effects.

The examination of safety claims could extend beyond such general points about non-lethal weapons. In considering the possible long-term effects of chemical weapons, no attention was given to the methodology for determining the toxicity, carcinogenicity or mutagenicity of chemicals, themselves matters where further considerations could be brought to bear.[92]

Something of the scope for continuing controversy about the benefits and risks of non-lethals can be illustrated by considering a different, more thoroughly regulated technology. The Himsworth Committee and members of the US Air Force Research Laboratory advise that non-lethal weapons should be subject to similar control procedures as pharmaceutical drugs. Leaving aside for the moment what this might mean in terms of specifics, John Abraham has argued that such licensing and regulating procedures in and of themselves do not guarantee agreement or fail-proof assessments.[93] Countries reach different conclusions about the safety of drugs.[94] Uncertainty is at the heart of many discussions and disagreements about drug regulation, uncertainty about the safety and efficacy of treatments, the scope of application of the findings of limited-scale trials, the criteria for trial inclusion, and the population that will ultimately take the medication. Cases of serious illness following drug use raise questions about whether those were caused by the drug, its interaction with other medications, various environmental exposures, etc.[95] Many adverse reactions are said to result from so-called 'off-label' practices, where patients or doctors do not follow the rules set out for a drug's prescription. Here, the problem rests with individuals rather than with the safety of manufacturers' products. A major source of concern for Abraham is the reliance of international regulatory bodies on industry-sponsored research. These points about drug regulation echo issues already discussed about non-lethal weapons and would suggest that disagreement and uncertainty are likely to continue for some time, whatever the testing regime.

One reason for the continuing scope for disagreement is that evidence about risk does not simply speak for itself. At a third level of the limitations of analysis, the processes for defining, identifying and interpreting health risks are open to question. The 'facts' of the matter are bound up with interpretations of what constitutes valuable and credible evidence. The examples of the FBI Firearms Training Unit tests and the safety claims of manufacturers illustrate something of the scope for credibility disputes. Questions about the credibility of evidence will come into play more in later chapters as specific deployments are examined in greater detail. At stake in questions about what counts as acceptable evidence are a host of social and technical issues that can only be reduced to simple 'facts' about 'what caused what' by a radical simplification of the issues at stake.

CONCLUSION

This chapter has taken as its starting point, and elaborated, how the effects of non-lethal weapons are the outcomes of social processes where evidence and resources are marshaled. What is known about this technology is inseparable from questions about how data are interpreted. To compare, for instance, the penetration depths of different kinetic-energy weapons into a gel block is not to resolve disputes about the acceptability; rather, it is a starting point for asking questions about how weapons will be used in practice, with what sorts of assumptions about their likely effects, in which situations, against whom, and in lieu of what other actions.

It has been suggested that many of the arguments about these weapons rely on equivocal evidence, where medical and technical forms of assessment have significant limitations. Like all assessments of weapons, this one suffers from some limitations. This chapter has only noted certain concerns and the scope for contending accounts of non-lethal weapons. Although it has problematized appeals to the objective facts of technology and brought up grounds for doubting certain general statements, it has not given much examination into what such points mean for the use of non-lethals, for their control, or for how praise or blame is determined in specific situations. It is to these matters that we now turn.

NOTES

1. See e.g. Wynne, B., in Rip, A., Misa, T. and Schot, J. (eds), 'Technology Assessment and Reflexive Social Learning', *Managing Technology in Society* (London: Pinter, 1995), 19–36; Nelkin, D. (ed.), *Controversies* (Newbury Park, CA: Sage, 1992); and Jasanoff, S., Markle, G., Peterson, J. and Pinch, T. (eds), *Handbook of Science and Technology Studies* (London: Sage, 1995).
2. Coupland, R., ' "Non-lethal" Weapons', *British Medical Journal*, 315 (1997), 72.
3. See Montgomery, Major N., 'Non-Lethal Weapons Human Effects', Presentation at Jane's Non-Lethal Weapons Conference, London, 1–2 November 1999.
4. Goolsby, T., 'Aqueous Foam as a Less-Than-Lethal Technology for Prison Applications', in Alexander, J., Spencer, D., Schmit, S. and Steele, B. (eds), *Proceeding of Security Systems and Nonlethal Technologies for Law Enforcement* (Boston: The International Society for Optical Engineering, 1997), 86–92.
5. Altmann, J., 'Non-Lethal Weapons', *Medicine, Conflict, and Survival*, 17 (2001), 234–47.
6. Alexander, J., 'Non-Lethal Weapons: The Generation After Next', in *Non-lethal Weapons*, Proceedings of the 1st European Symposium on Non-Lethal Weapons, 25–26 September 2001, Ettlingen, Germany (Postfach: ICT, 2001).
7. Omega Foundation, *Crowd Control Technologies*, Report to the Scientific and Technological Options Assessment of the European Parliament, PE 168.394 (Luxembourg: European Parliament, 2000).
8. Murphy, M., Jauchem, J., Merritt, J., Sheery, C., Cook, M. and Brown, G., 'Acoustic Bioeffects Research for Non-Lethal Applications', in *Non-lethal Weapons*, Proceedings of the 1st European Symposium on Non-Lethal Weapons 25-26 September 2001 Ettlingen, Germany (Postfach: ICT, 2001), 9–8 to 9–9.
9. A.L.S. Technology, 'Gunsmith Learned his Trade being one of America's Best', Bull Shoals, Arkansas: A.L.S. Technology [cited 5 January 2002]. See www.ozarkmtns.com/less-lethal/dave.htm
10. See e.g. Ijames, S., 'Less-Lethal Projectiles', *The Tactical Edge*, Fall (1996),76–84; Ijames, S., 'Less-Lethal Options', *The Police Chief* (1997), 31–7; Ijames, S., 'Emerging Trends in a Hot Realm', *The Tactical Edge*, Winter (1999), 71–3.
11. Ijames, S., 'Testing and Evaluating Less-Lethal Projectiles', *The Tactical Edge*, Spring (1997), 12–15, and Ijames, S., 'Testing and Evaluating Less-Lethal Projectiles- Part 2', *The Tactical Edge*, Fall (1997), 87–92.
12. Specifically, kinetic energy equals one-half mass times velocity squared, or $KE=mv^2/2$. See Egnar, C., *Modelling for Less-lethal Chemical Devices*, US Army Engineering Laboratory Technical Report.
13. Omega Foundation, *Crowd Control Technologies*, 57.
14. Cuadros, J., 'Definition of the Lethality Thresholds for KE Less Lethal Projectiles', in Alexander et al., *Proceeding of Security Systems*, 20–7.
15. See Widder, J., 'Review of Methodologies for Assessing the Blunt Trauma Potential of Free Flying Projectiles Used in Non-Lethal Weapons', in *Proceedings of NDIA Non-Lethal Defense IV Conference*, 20–22 March 2000 (National Defense Industrial Association Conference) [cited 15 November 2001]. See www.dtic.mil/ndia/nld4/index.html. See as well Defense Technology and Federal Laboratories, *Research Journal of Health Risk Evaluations for Less Lethal Products* (Casper, WO: DT/FL, n.d.).
16. Zitrin, R., 'Cops Probe Death of Man Shot with Beanbags', *APB News*, 20 May 1999. See www.apbnews.com/.
17. Hubbs, K., 'Less Lethal Munitions and Civil Disturbance Responses', *The Tactical Edge*, Spring (1997), 9–11.
18. See Thomas, T., Grange, J. and Linda, L., 'Injuries Associated with Less Lethal Weapons', *Lifeline*, 12 December 2000. See www.calacep.org/lifeline/displaylifeline.html?ID=135
19. De Brito, D., Challoner, K., Sehgal, A. and Mallon, W., 'The Injury Pattern of a New Law Enforcement Weapon', *Annals of Emergency Medicine*, October, 38, 4 (2001), 383–90.
20. Sehgal, A. and Challoner, K., 'The Flexible Baton TM-12', *Journal of Emergency Medicine* 15, 6 (1997), 789–91; and Mangold, T., *Panorama – Lethal Force*, aired BBC, Sunday 9 December 2001; Blue Line, 'Force Agrees to Stop using Weapon', *Blue Line – Canada's*

National Law Enforcement Magazine, April 1997.

21. De Brito, D. et al., 'The Injury Pattern'.
22. Shank, E. et al., *A Comparison of Various Non-Lethal Projectiles* (Aberdeen, MD: US Army Land Warfare Laboratory, 1974).
23. See e.g. Ijames, 'Testing and Evaluating'.
24. UK Steering Group for Patten Report Recommendations 69 and 70, *A Research Programme into Alternative Policing Approaches Towards the Management of Conflict* (Belfast: Northern Ireland Office, 2001).
25. Kenny, J., Heal, C. and Grossman, M., *The Attribute-Based Evaluation (ABE) of Less-Than-Lethal, Extended-Range, Impact Munitions* (2001). See www.arl.psu.edu/areas/defensetech/gifs/abe_report.pdf
26. Kendig, T., 'Non-Lethal Weapons Testing Proves Many Inaccurate', *Penn State Intercom*, 24 May (2001). See www.psu.edu/ur/archives/intercom_2001/May24/weapons.html
27. Kenny et al., *Attribute-Based Evaluation (ABE)*.
28. Kendig, 'Non-Lethal Weapons Testing'.
29. Miller, R., Rutherford, W., Johnston, S. and Malhorta, V., 'Injuries Caused by Rubber Bullets', *British Journal of Surgery*, 62 (1975), 480–6, and Rocke, L., 'Injuries Caused by Plastic Bullets Compared with those Caused by Rubber Bullets', *The Lancet*, 23 April (1983), 919–20.
30. UK Steering Group, *A Research Programme*, 46.
31. See Committee on the Administration of Justice, *The Misrule of Law*, October (Belfast: CAJ, 1996), and Omega Foundation, *Crowd Control Technologies*, 30.
32. Waddington, P. A. J., *The Strong Arm of the Law* (Oxford: Clarendon, 1991), 205.
33. Committee on the Administration of Justice, *Plastic Bullets*, June (Belfast: CAJ, 1998) and Omega Foundation, *Crowd Control Technologies*, 30. It was not until after various campaigning activities by concerned groups that in 1997 the guidelines given to officers for the bullets were made public. Until the rules were revised in 2000, plastic bullets were justified by the need to protect life or property, preserve the peace, and prevent or detect crime.
34. Defence Scientific Advisory Council, 'Statement on the Comparative Injury Potential of L5A7 Baton Round Fired from the L104 Anti-riot Gun using the Battle-Sights, and the L21A1 Baton round fired Using the XL18E3 Optical Sight', 21 August 2001.
35. Wright, S. and Evans, R., 'A Shot in the Dark', *The Guardian*, 28 June 2001.
36. Straw, J., 'Home Department: Baton Rounds', *Hansard*, 2 April 2001.
37. Evans, R., *Gassed* (London: House of Stratus, 2000), Chapter 9.
38. Ibid.
39. Pear, R., 'Cancer in DC Police Tied to Tear Gas Use', *Washington Star*, 31 October 1972, 1.
40. Dyer, Robert F., *Personal Letter to Robert R. Jones*, n.d.
41. *Home Office Report of the Enquiry into the Medical and Toxicological Aspects of CS (Orthochlorobenzylidene Malononitrile)* (London: HMSO, 1971), 48.
42. See Ackroyd, C., Margolis, K., Rosenhead, J. and Shallice, T., *The Technology of Political Control* (London: Pluto, 1980).
43. For a typical statement of these claims see Edwards, S., Granfield, J. and Onnen, J., *Evaluation of Pepper Spray* (Washington, DC: National Institute of Justice, 1997).
44. Salem, H., Olajos, E., Miller, L. and Thomson, S., *Capsaicin Toxicology Review* ERDEC-TR-199 (US Army Edgewood Research, Development, and Engineering Center, 1994).
45. American Civil Liberties Union, *Oleoresin Capsicum*, June 1995.
46. Doubet, M., *The Medical Implications of OC Spray* (Millstadt, IL: PPCT Management Systems, 1997).
47. Smith, M. and Alpert, G., 'Pepper Spray', *Policing*, 23, 2 (2000), 242.
48. For a listing of countries see Steering Group, *A Research Programme*, 49–50.
49. Busker, R.W. and van Helden, H., 'Toxicological Evaluation of Pepperspray as a Possible Weapon for the Dutch Police Force', *The American Journal of Forensic Medicine and Pathology*, 19, 4 (1998), 309–16.
50. Ibid., 315.
51. See Watson, W., Strmel, K. and Westdrop, E., 'Oleoresin Capsicum (Cap-Stun) Toxicity

from aerosol Exposure', *Annuals of Pharmacother* (1996), 733–5.
52. Granfield, J., Onnen, J. and Petty, C., *Pepper Spray and In-Custody Deaths*, Executive Brief (Alexandria, VA: International Association of Chiefs of Police, 1994). Throughout this chapter, comments about this study are based on this 'Executive Brief'. A conversation with the International Association of Chiefs of Police [21 January 2002] indicated no full report was produced. In any case, this brief has served as the source for those mentioned in this chapter who have drawn on the study to advance particular claims.
53. Busker, R., van de Meent, D. and Bergers, W., 'Safety Evaluation of Pepperspray in the Ovalbumin Sensitized Guinea Pig', in *Non-lethal Weapons*, Proceedings of the 1st European Symposium on Non-Lethal Weapons, 25–26 September 2001, Ettlingen, Germany (Postfach: ICT, 2001).
54. Personal Communication, Ruud Busker with author, 18 April 2002.
55. US Department of Justice, News Release, 12 February (Miami: United States Attorney, Southern District of Florida, 1996).
56. E.g. Pilant, L., *Less-than-Lethal Weapons* (Washington, DC: International Association of Chiefs of Police, 1993).
57. Longman, C., *Cap-Stun Weapons Systems: Aerosol Product Line* (Bethesda, MA: Zarc International, 1993). Both the claims that ZARC and the US Army Chemical Research and Development Center were involved in the testing process have been challenged elsewhere; see Doubet, *The Medical Implications of OC Spray*.
58. Shapiro, H., 'Letter to Alan Parachini of the ACLU Foundation of Southern California', 17 May 1996 [cited 5 January 2002]. See www.zarc.com/english/news/fbiaclu.html
59. Doubet, *The Medical Implications of OC Spray*.
60. Rogers, T. and Johnson, S., 'Less than Lethal', *International Journal of Police Science and Management*, 3, 1 (2000), 55–67.
61. SEASKATE, *The Evolution and Development of Police Technology* (Washington, DC: SEASKATE, 1998).
62. Los Angeles Times, *Los Angeles Times*, 18 June 1995.
63. Doubet, *The Medical Implications of OC Spray*, 27.
64. Kaminski, R., Edwards, S. and Johnson, J., 'Assessing the Incapacitative Effects of Pepper Spray During Resistive Encounters with the Police', *Policing*, 22, 1 (1999), 7–29; Lumb, R. and Friday, P., 'Impact of Pepper Spray Availability on Police Officer Use-of-Force Decisions', *Policing*, 20, 1 (1997), 136–48; Rogers and Johnson, 'Less than Lethal'; SEASKATE, *The Evolution and Development of Police Technology*; and Smith and Alpert, 'Pepper Spray'.
65. Chan T., Vilke G., Neuman T. and Clausen, J., 'Restraint Position and Positional Asphyxia', *Annals of Emergency Medicine*, 30, 5 (1997), 578–86. For a description of some of these issues see 'Startling News on Positional Asphyxia; Effects of the "Hog-Tie" Called into Question; Dr. Reay Changes Opinion', November home.earthlink.net/~greg-meyer/articles/update.html
66. Steffee, C., Lantz, P., Flannagan, L., Thompson, R. and Jason, D., 'Oleoresin Capsicum (Pepper) Spray and "In-Custody Deaths"', *The American Journal of Forensic Medicine and Pathology*, 16, 3 (1995), 185–92.
67. Ibid., 191.
68. Chan, T., Vilke, G., Clausen, J., Clark, R., Schmidt, P., Snowden, T. and Neuman, T., *Pepper Spray's Effects on a Suspect's Ability to Breathe*, December (Washington, DC: National Institutes of Justice, 2001), 1.
69. The 1997 work carried out by Chan and colleagues that refuted the negative effects of 'hog-tying' did not test for these conditions either.
70. International Association of Chiefs of Police, *Pepper Sprays Evaluation Project* (Washington, DC: National Institute of Justice, 1995) and Edwards et al., *Evaluation of Pepper Spray*.
71. Kaminski et al., 'Assessing the Incapacitative Effects'.
72. Ibid., 19.
73. Berkeley Police Review Commission, *The Effectiveness of Pepper Spray* (Berkeley, CA: Berkeley Police Review Commission, 1997).
74. Messina, P., 'Chemical Deterrent Sprays', *Law Enforcement Technology*, October 1993, 106–8.

75. Kaminski et al., 'Assessing the Incapacitative Effects'.
76. Smith, C. and Stopford, W., 'Health Hazards of Pepper Spray', *North Carolina Medical Journal*, 60, 5 (1999), September, 268–74. See www.ncmedicaljournal.com
77. Shapiro, H., 'Letter to Alan Parachini of the ACLU Foundation of Southern California', 17 May 1996.
78. For a report of allegations of this see Amnesty International, *Pepper Spray Used Repeatedly on Native American Children*, AMR 51/049/2001 (London: Amnesty International, 2001).
79. Smith and Stopford, 'Health Hazards of Pepper Spray'.
80. Kaminski et al., 'Assessing the Incapacitative Effects', 12.
81. Baltimore County Police Department, *Baltimore County Police Department Lesson Plan – OC Aerosol Spray* (Baltimore, MD: BCPD, 1993), 10.
82. Smith and Stopford, 'Health Hazards of Pepper Spray'.
83. Rhodes, N., 'Pepper Spray, Product Liability and Cops', *Policing by Consent* (1996),12–13.
84. See Defense Technology and Federal Laboratories, *Research Journal of Health Risk Evaluations for Less Lethal Products* (Casper, WO: DT/FL, n.d.).
85. Doubet, *The Medical Implications of OC Spray* and Rhodes, 'Pepper Spray, Product Liability and Cops', 12–13.
86. Murphy et al., 'Acoustic Bioeffects Research for Non-Lethal Applications'.
87. Merrick, J., 'Magic Bullets They're Not', *American Police Beat*, 1, 5 (1994), 1.
88. Goodrich, J., *Letter to Customers from Mace Security International* (Bennington, VT: Mace Security International, n.d.).
89. Lindblom, C. and Woodhouse, E., *The Policy-Making Process* (Englewood Cliffs, NJ: Prentice-Hall, 1993). See as well Sarewitz, D., *Frontiers of Illusion* (Philadelphia: Temple University Press, 1996).
90. Woodhouse, E., 'Is Large-Scale Military R&D Defensible Theoretically?', *Science, Technology, and Human Values*, 15, 4 (1990), 442–60.
91. Vasel, E. and Marrero, J., 'New Horizons', 2002. See http://www.dtic.mil/ndia/2002nonlethdef/Vassel.pdf
92. For one such discussion see Thornton, J., *Pandora's Poison* (Cambridge, MA and London: MIT Press, 2000).
93. Abraham, J., *Science, Politics and the Pharmaceutical Industry* (London: UCL Press, 1995) and Abraham, J., 'Regulating the Cancer-Inducing Potential of Non-Steroidal Anti-inflammatory Drugs', *Social Studies of Medicine*, 46 (1998), 39–51.
94. Abraham, J. and Sheppard, J., 'Complacent and Conflicting Scientific Expertise in British and American Drug Regulation', *Social Studies of Science*, 29, 6 (1999), 803–43.
95. As an indication of this, those in the pharmaceutical industry and elsewhere often refer to side effects from medicines as 'adverse drug events' rather than 'adverse drug reactions' in order to question any casual link.

6

On to the Streets: Examining Major Deployments of Non-Lethals

The previous chapter traced out the basis of some health and safety statements made about certain prominent and well-established non-lethal weapons. In doing so, various inconsistencies, contingencies and assumptions were said to be at play in safety claims. Just what the risks might be, to what extent they are acknowledged, and how seriously they should be taken are matters where contending accounts are on offer. It was suggested that many non-lethals are situated in a murky area where claims about what these technologies do and do not do are contested and contestable. Notions of just what a weapon is are not merely reflections of some properties read off from manufacturers' catalogues, but they are inseparable from matters of interpretation.

This chapter extends the analysis of the legitimacy of non-lethals by examining topics associated with their deployment. Chapter 5 already illustrated something of how medical assessments rely on and inform assumptions about the manner in which weapons are used. Finding out how claims about the 'actual effects' of non-lethals are advanced requires considering the situational use of these devices. Particular attention in this chapter is given to the rules governing non-lethals and the contexts of the weapons' use. Both provide a further grounding for thinking through questions about legitimacy. First, drawing on various academic analyses, attention is given to the potential of using force according to rules and deviations from rules. Second, the last chapter examined non-lethals primarily in relation to either unspecified contexts or a limited range of scenarios. This chapter expands the range of contexts. In doing so, 'context' itself becomes a topic for analysis and debate. Just as the last chapter questioned definitive accounts of 'just what a technology does', so this one illustrates the scope for negotiation over what context is at stake. The aim is to consider how certain claims about non-lethals are said to be true and how their use is considered legitimate or illegitimate. What is at stake is how casual judgments are formed and how these then help form determinations of responsibility.

OPERATING BY THE RULES

As outlined previously, ensuring non-lethal weapons cause minimum harm requires following certain rules or abiding by certain practices. Thus, the legitimacy of the recourse to non-lethals, in part, depends on whether actions conform to rules and whether those rules are justified. Possible proscriptions include considerations about who is targeted, in which part of the body, from what distance, for what duration, and with what sort of aftercare. Rules also provide the basis for safety claims. So in the case of incapacitant sprays, their effects might be considered relatively non-dangerous given that they are activated in a particular way (one short-term burst) and from a certain distance. And yet, as suggested previously, the feasibility of acting in such a manner is not something that necessarily informs technical or medical assessments. Whether or not one maintains that rules are being adhered to in turn affects assessments of the cause of injuries. Where the rules are followed and yet significant injury results, the weapons in question may be deemed more dangerous than originally thought.

Given these considerations, significant importance is attached to whether or not operational rules have been abided by. Initially it can be noted that there are instances in which it is widely agreed that rules are not followed. Non-lethals can be used inappropriately or abusively. For tear gas to act as one of the most effective and humane weapons, as suggested by the British Secretary of State in Chapter 3, requires that individuals be able to flee the gassed area. Sadly, this has not always been the case. In May 2001, more than 130 people died in a stampede at a football game in Ghana when multiple tear-gas canisters were fired into the crowd.[1] In this case, the escape exits had been closed and individuals were crushed as they tried to flee.[2]

Alternatively, such weapons might be used in ways identified as less than humane in intent. In past years, organizations such as Amnesty International have contended that various forms of riot-control equipment have served as tools of political repression.[3] Take the case of past practices of security forces in Kenya: there, concerns were voiced about both the excessive and inappropriate use of tear gas.[4] Kenyan human-rights activist, Janai Robert Orina, described conditions in the capital,

Nairobi, during the late 1990s with the following terms:

> Tear gas is a day-to-day experience for us ... There are times when the air around the city of Nairobi reeks of it ... In the confusion when we are blinded and convulsing from the gas, the police move in to club and beat us.

One such incident took place in July 1997. Amnesty reports that the Kenyan paramilitary police stormed an Anglican Cathedral in the capital after pro-reform activists took refuge there after a peaceful political protest was dispersed. Security forces threw tear-gas canisters inside the cathedral and then moved in wielding truncheons. In such an enclosed space, tear gas functions as a tool more for crowd punishment than for crowd control.

Rule Deviation as Normal?

While few would doubt that weapons can be used improperly, in democratic countries with highly trained police and military forces such practices are supposed to be kept to a minimum. To what extent, though, are actions described as rule-governed?

Much of criminological research about the use of force by the police would question whether the conduct of officers should be understood as governed by rules; this, at least, in the sense of comprehensive do's and don'ts. Hunt, for instance, echoes a common argument in contending that a noticeable divide exists between the formal rules regarding the use of force and the informal practices that characterize policing.[5] While officers might be taught to target certain parts of the body to minimize harm, in volatile and confused contexts such prescriptions often go out in favor of using the force seen as necessary in a given situation.[6] For Hunt and many others, such statements are not meant to vilify, but attest to the dilemma of force in democratic societies. Similarly, this does not imply that any force can be justified as reasonable or that training cannot encourage certain practices. Rather, these conclusions attempt to recognize the difficulty of devising general rules about conduct that are supposed to apply to particular real-world situations. As she summarizes:

> The organization of policework reflects a poignant moral dilemma: for a variety of reasons, society mandates to the police the right to use force but provides little direction as to its proper use in specific, 'real life' situations. Thus, the police, as officers of the law, must be prepared to use force

124

under circumstances in which its rationale is often morally, legally, and practically ambiguous.[7]

High-profile cases such as the beating of Rodney King illustrate the potential for different standards across society as to what constitutes acceptable force. Alternative lay, administrative, professional, and legal standards exist about what counts as reasonable force (or 'proportionality' in military terms), what rules can and should be followed and how closely.[8]

Such conclusions are complemented by more general studies of policing.[9] While the relation between formal rules and practices is less than straightforward in any organization, it is often said to be particularly problematic within the police. Despite the link between policing and law enforcement, discretion in the adherence to rules is central to the actions of police officers.[10] Because the number of law-breaking incidents is far too numerous to permit taking actions against all of them, the police must make judgments about the interpretation of rules and decide how to enforce the law in practice. While in the public 'frontstage', police are presented as quasi-militaristic and highly disciplined; a central theme of criminology is that 'backstage' on the streets, that image gives way to a much more messy set of practices.[11] Further compounding the situation, most day-to-day policing has a 'low visibility' in relation to monitoring activities of police management and the general public. Attempts to control the actions of rank-and-file officers, or even accurately represent them, often founder for this reason.[12]

The use of force is one of those areas often (but not always) characterized by a low visibility, where discretion exists about when and what action needs to be taken. In countries such as the United States and Britain, the law states that force must be reasonable and necessary.[13] Just what these terms mean in relation to concrete contexts is a matter for negotiation. There is no universally valid and unconditional list of steps for officers to follow with regard to the use of force. A person immersed in a conflict situation involving anger or fear is likely to be less dispassionate than someone reading a book. Notions of what constitutes reasonable actions are typically evoked in relation to the perspective of individuals at the scene rather than with hindsight.[14] So while use-of-force guidelines might state that certain acts are only justified when there is a threat of danger to members of the public or the police, just what constitutes a threat is highly contingent upon the situation in question. As one author put it, 'What is defined as unreasonable on the day depends on what has gone on before. What is reacted to violently in one setting, with particular structural, political/ideological, cultural, contextual and situational features, may evoke a less dramatic response in another.'[15] Along these lines, Ericson and Doyle, for instance, argue that in the context of the policing of international political events, the resort to blatantly undemocratic tactics by security forces is ever possible because of the global repercussions of being seen as unable to maintain order.[16] While discretion enables a flexibility that resists a mechanical approach to the use of force, it also brings grounds for criticism for decisions by those outside the police and the potential for conflict between rank-and-file and senior officers.

What the authors cited above suggests is that the use of force by the police does not take place in the proscribed and specific ways according to neat rules. Statements to such an effect might have a utility in public relations, but they are arguably not a basis for speaking about the likely practical applications. As mentioned in Chapter 5, despite rules about the appropriate targeting distance for using pepper sprays and kinetic-energy weapons, there is evidence that such rules are broken with some frequency. Such deviances are justified, though, in relation to the perspective of the perceived threat in that specific context and of the corresponding action deemed necessary.

More than just violating given rules for a given situation, the argument has been made that the availability of certain weapons can help redefine particular circumstances by transforming the choices, capabilities and experiences of individuals. Lumb and Friday, for instance, evaluated a trial of OC sprays. While maintaining that OC sprays reduced the likelihood of injury in specific instances, they contended that the devices also gave officers a greater confidence to be able to intervene in situations through the use of force, thus expanding the range of force incidents. As they concluded, 'the use of spray may not necessarily be an alternative to [physical or deadly] force, but provides officers with options to use more force – perhaps unnecessarily. In other words, if it is there, they will use it.'[17] Thus, the interpretation of the context in question and what action needed to be taken are established in relation to the understanding of a technology (here, that the OC sprays represented a valuable and relatively benign option).[18]

Those who study 'risky technologies' echo the limitations of understanding actions as rule-governed. Wynne has commented that discussions about many technologies are often characterized by a stark divide between the rule-bound images in public discussions and the contingencies and informal routines (even rule-breaking practices) prevalent in actions.[19] Safety assessment and maintenance procedures for aircraft, for instance, are done in relation to how commonplace particular faults are, how these can be compensated for, and how realistic it would be to conduct thorough evaluations and repairs of problems. In everyday life, taking something as mundane as driving a car – in relation to driving within speed limits, checking tire pressure, and attending to any abnormal noises through diagnosis and repair – gives ample illustrations of the way individuals routinely do not use technologies according to strict rules. These statements are not made to condone any actions. For Wynne, experts face many dilemmas as they try to reconcile the rule-bounded images of technology with the messy world of practice. Often, this means that for public audiences, accidents or other problems are attributed to 'human error' or deviations from rules by operators, even when the experts involved recognize the limitations of formalized rules in reflecting the operational practices.

This section has suggested that reasons exist to believe that the adherence to rules regarding non-lethal weapons is not a straightforward matter. Not only is it naive to believe that the use of force necessarily takes place in accordance with rules, such a view ignores how individuals define and respond to particular situations and thus what action should be taken. It is not being suggested that the points above imply that training or rules of engagement for the use of force are unnecessary or useless. Both are key areas for consideration. Rather, I initially want to note that the adherence or lack of it to rules is likely to be a matter where there is some debate. Once this is acknowledged, the question of whether or not rules are being followed gives way to other questions: How does the scope for discretion become defined within organizations? How are assessments about the adherence to rules tied in with determinations of where responsibility rests for ensuring the appropriate use of weapons? These issues are taken up later. Next, though, let us pick up on a number of points alluded to in this section about 'context'.

REDEFINING CONTEXTS

It seems reasonable to suggest that determinations of the appropriateness of deploying non-lethals vary by the situation. Where the police attempt to break up domestic disputes or disarm suicidal persons, there is likely to be little disagreement that something needs to be done. Whether and when the resort to a weapon, albeit a non-lethal one, is the best course of action is another matter. Yet, it would seem reasonable to say that such incidents are testing one, where non-lethals might provide valuable options. Likewise in cases where the lives of security personnel are threatened, say by an attacker with a knife or some such weapon, the possible utility of non-lethals is likely to be greatest.

Just what context a technology should be evaluated against, though, is not something that remains fixed. A vivid account of this is given by the use of CS gas during the 1960s by the US Army in Vietnam. Initially, the stated expectation was that the irritant would be employed 'only in those situations involving riot control or situations analogous to riot control'.[20] In such limited circumstances, the gas would function much the same as tear gas in the United States at the time. After some initial deployments, the remit expanded to include the separation of combatants from non-combatants. As Dando argues, though, as commanders became familiar with the technology, its range of uses expanded considerably.[21] Between 1964 and 1970, nearly 16 million pounds of CS was procured for operations in Vietnam. The range of field uses included forcing Vietnamese out of underground hideouts (perhaps then followed by bombing runs from B-52s[22]), making areas uninhabitable, and (as in the First World War with other gases) harassing enemy soldiers in combat. According to US Army reports, the possibility of employing CS to separate combatants from non-combatants proved elusive, though, as civilians typically fled combat areas before fighting began.[23]

Such a shifting set of situations illustrates how the aims for weapons can develop over time. As with many other technologies, the functions of non-lethal weapons are not settled beforehand, but actively determined in the course of their use.[24] Aims and means undergo progressive clarification and elaboration during the deployment. Each alteration in turn brings the possibility of reassessing the merits of such technology.

Yet, possible reassessments go beyond just re-evaluating the safety or effectiveness in light of a weapon's use in a new situation. In England and Wales, CS sprays were introduced in 1996 as a means for officers to protect themselves or members of the public from physical threat. When such devices began to be used in psychiatric wards and other healthcare facilities, their acceptability came into doubt. In relation to the spray's solvent, fears were raised about the potential effects for mentally ill individuals taking neuroleptic drugs. More fundamentally, though, questions were raised about the appropriateness of targeting the mentally ill. In 1998, based on a survey of mental-health units, Bell and Thomas argued that the sprays were 'totally at odds with the therapeutic role of health care professionals'.[25] One mental-health director surveyed summarized many of the issues identified in suggesting:

> I subscribed to the view that the people who need to come into our wards are mentally ill, not criminal, and that means they should not be sprayed with the gas. Before the police would have had to spend more time talking to the patients, even if maybe in the end they had to resort to physical force. This spray seems to me like a short cut; a quick and easy answer. It's yet another sign of the way criminality and

mental illness get conflated, just when we are working so hard to change public attitudes ... I fully support the police using the gas when faced by violent criminals, but these are people our nurses on the acute admissions wards deal with on a daily basis, and they don't have these gadgets available to them.[26]

This statement suggests various issues about how the appropriateness of non-lethals might vary by context, how the use of certain technologies helps define the identity of individuals in that setting (as criminals), how responses to situations by the police might alter because of the availability of certain options (by the quicker resort to force), and how alternative responses are undertaken (by nurses, every day).

The possibility that the remit of particular non-lethal weapons might gradually expand or contract over time in ways judged as inappropriate is tied in with wider questions about the operation of security organizations. As discussed in Chapter 4, the uptake of new forms of weapons in Western police forces is often linked to their 'para-militarization'. In the United States, Kraska and Kappeler have expressed concerns about a growing paramilitarism.[27] As they contend, since the 1970s, SWAT and other elite paramilitary units moved from being a small part of some police departments to an integral part of nearly all major agencies. This spread in paramilitary units has taken place alongside their growing utilization. Between 1983 and 1995, the number of call-outs for paramilitary units quadrupled. While in the past these units responded to a limited range of situations, such as riots or hostage-taking, as part of the 'War on Drugs' that remit has shifted toward proactive policing of even minor offences. Paramilitary units are becoming normalized into routine police work such as patrolling certain 'hot spot' areas. Situations in the past that would have been handled by standard forms of community policing have become the domain of specialized units. The primary danger that Kraska and Kappeler identified with such units was the increasing practice of viewing complex social problems as resolvable through military-style force: a sentiment shared elsewhere.[28] As part of this has come the further deployment of weaponry in the use of force, including non-lethals originally justified for a narrow range of situations. Given these practices and the elite status of paramilitary units within police departments, the units act as both testers and promoters of new technologies.[29]

Such sentiments have been countered elsewhere. P. A. J. Waddington maintains that what separates paramilitary units from other police officers is not that they are more aggressive, but that they are better trained and thus better able to resolve situations without serious harm. This might entail the use or display of force, but the overall result is a reduction in injury to the police and the public.

Within such debates there are contrasting ways of understanding how organizations employ weapons. In many critical accounts of para-militarism, the functioning of non-lethal weapons and other equipment reflects the overall social and organizational practices in the police. The importance given to both adopting military-style paraphernalia and command structures along with the discretion of policing provides the basis for an expanding use of such technologies in aggressive operations. For Kraska and Kappeler, the implications of the adoption of weapons also extend beyond the situations of their deployment. Symbolically, the greater eagerness to resort to non-lethal weapons, associated equipment and tactics reinforce the belief in the appropriateness of force. In contrast, authors such as Waddington choose to frame the important issues about weapons in terms of their stated goals and effects. As security forces must respond to difficult situations, they require appropriate force options.

PUBLIC ORDER AND THE ORDERING OF THE PUBLIC

Concerns about the varied appropriateness of non-lethal weapons across contexts, the symbolic aspects of actions, and the possibility that a response to situations might alter because of the availability of certain options become stark in discussions about the policing of civilian crowds. Here, questions about the legitimacy of force tie in with assumptions about the identity of those taking part in 'public order' events and about their motivations.

A prominent example of the possibility for alternative, sometimes diametrically opposed, versions of events about the appropriateness of the resort to non-lethal weapons is provided by the case of the World Trade Organization (WTO) conference in Seattle during November and December of 1999. Images spread around the world of confrontations between police clad in all-black protective uniforms engaging with crowds under a white-yellow haze of tear gas. Many of the most dramatic such scenes took place around key sites of the conference, as protesters attempted to resist access to buildings and the police attempted to clear the streets. Various non-lethal or less-lethal weapons were deployed, including tear gas, rubber bullets and handheld pepper sprays. While the events of those days have been subject to considerable discussion,[30] and no doubt will be subjected to more in the future, here I only want to consider a limited range of issues related to the use of non-lethal force.

Criticisms of the police actions included allegations from human rights organizations, civil liberty groups and others regarding the indiscriminate use of non-lethals against non-violent protesters and bystanders.[31] Suggestions that orders to disperse were lacking or unclear meant that many individuals were targeted improperly, including those fleeing from areas. Vaguely defined 'no-go' protest zones were established around certain sites that meant anyone in them became open for arrest and the use of force. Allen argues that OC and CS chemical sprays and gases of questionable safety in best-case situations were used against their guidelines and affected people with heart conditions, asthma, diabetes and other ailments that made them particularly vulnerable.[32] The overall failure of the police to provide aftercare facilities, such as water to treat eyes sprayed with OC, meant that the effects experienced were much more severe than they should have been. The Seattle National Lawyers Guild said that non-lethal kinetic-energy projectiles and grenades were used at a distance closer than that recommended by manufacturers. The use of non-lethals such as large chemical dispensers, though highly visible, meant the pain experienced by protesters was ungaugeable and thus more palatable to the police and the public.[33] Well after the protest, anecdotal stories have been given of lasting ill effects.

Members of the Seattle Police Department offered a sharply contrasting account in their After Action Report.[34] A summary of the riots concluded:

> The professionalism and restraint displayed by the police officers, supervisors, and commanders on the 'front line,' whether posted at venues or assigned to demonstration management or escort duty, was nothing short of outstanding. This review of the WTO event found nothing to rival this single point: without the remarkable poise and performance of front line officers and their supervisors, the WTO Conference event could have concluded far differently than it did.

Despite this overall conclusion, certain weaknesses were identified in the department's practices. The report noted a pre-event failure to assess the strengths of a 'well-trained and equipped adversary'. The Department's preparation had been based on past experience with peaceful protesters instead of the new paradigm of disruptive rioters as represented by those at Seattle. The resulting failure to plan adequately for a worst-case scenario left the police operationally ill-prepared. Procedures for re-equipping officers who ran low in, or out of, CS gas and OC sprays, or for supplying officers with enough protective equipment, were not thought through properly. On several occasions, the command and control structures broke down and officers on the ground were not given sufficient support.

In response to the WTO conference, the Seattle City Council under-took its own accountability review by conducting some 200 interviews.[35] The report of that review process concluded:

> We find that city government failed its citizens through care-less and naïve planning, poor communication of its plans and procedures, confused and indecisive police leadership, and imposition of civil emergency measures in questionable ways. As authorities lost control of the streets they resorted to methods that sometimes compromised the civil rights of citi-zens and often provoked further disturbance.[36]

In relation to the question 'compared to what?', alternative strategies for dealing with the disturbances on the days of the conference were outlined:

> If the goal was to clear the streets without inflicting death or widespread serious injury, police operations succeeded. But police operations must also have as their goal the protection of lawful speech and assembly, and should convey a percep-tion of even-handed commitment to protecting demonstra-tors as well as the larger public. Seattle used tear gas and other less-lethal technologies when large-scale arrests (which were expected by many demonstrators) would better have served the goal of restoring order.[37]

The use of chemical irritants was deemed particularly unsuitable because of the effects they caused in protesters, residents and bystanders alike. The review committee took a mixed position in relation to the responsibility of officers for the unfortunate outcomes. On the one hand, rank-and-file officers were poorly supported (as the police department review also maintained), but there were also individual acts of 'gratuitous assaults'. In some cases, officers failed to follow the proper rules with non-lethal weapons, such as giving a warning shout to disperse before their use, because they were under-staffed and exhausted. The police were let down by the failure to adequately staff the events so that necessary numbers were available to clear the streets by arresting individuals. Finally, the conclusions of the review were made conditional in noting:

The committee was disturbed at the slow and evasive production of records and documents by police commanders during our inquiry. Our investigators and citizen panels had constant difficulty getting information. [One of the panels] observed that 'whether due to poor internal management and disorganization, or deliberate efforts to withhold documents, neither explanation promotes confidence in the senior leadership of the department'.[38]

These evaluations of the review committee diverge significantly from the finding of the police department. While both agree on some points, such as deleterious effects of the loss of command and control structures, their overall findings are at odds. It might initially be thought that the findings of the city council's review committee would be warmly welcomed by those mentioned above who criticized the police use of force. But even among those who share a critical assessment of the overall handling of the conference, the findings of the city's review committee have been called into doubt. Many of the reasons for this relate to the issues regarding the definition, identification and interpretation of the facts and context. The Committee for Local Government Accountability (CLGA) in Seattle challenged the review's conclusions by citing a number of factors that called into question the evidential basis of the conclusions. These included the confiscation of recording devices by the police, the targeting of some members of the press with non-lethals to prohibit them from documenting the full range of police practices, and the failure of many officers to display a visible identification.[39] CLGA contended that the police misuse of force was widespread rather than limited, a claim which could have been substantiated more forcefully had the actions above not been taken. The credibility and independence of the city council's review committee was itself called into doubt as various roadblocks were said to restrict the input of the views of members of the community into the report. The set of grievances raised could go on. In short, though, seemingly pertinent doubts were raised about the factual basis of the review.

Such issues are important because even if one accepts that there were inappropriate police practices, as CLGA and the city council review do, just why these took place is not something widely agreed upon. Determinations and evaluations of what really happened would inform and be informed by assessments about who or what was responsible for certain outcomes. In the accounts given above, for instance, the source of the problems with injuries and excessive force identified related to sources as varied as the unsubstantiated safety claims of manufacturers, the unacceptable variability of weapons across the population, the callous actions of rank-and-file officers, and the lack of police training and preparation. The city council review casts officers as both victims and villains to the fray. While officers could have acted more appropriately, they were also under pressures that placed limitations on what could reasonably be expected.

As argued in the previous section, key concerns exist about whether the use of force in general by the police should or could be understood as rule-bound. In situations of the policing of crowds, such concerns are arguably particularly acute. Social psychologist Stephen Reicher, for instance, suggests that the policing of crowds raises quite distinctive practical problems for security personnel.[40] The difficulty of picking out who in the crowd was responsible for, or might initiate, attacks on police or property, the orientation of police toward the crowd as a whole (for instance in clearing the streets) rather than particular individuals, and the inability of police to be able to differentiate between members of the crowd[41] all raise problems about the likely adherence to detailed and specific rules. Although officers might recognize the diversity of motivations and behaviors of individuals within a crowd, acting on this is problematic. Once the crowd is handled as a group, members of it further identify themselves as part of a group opposed to the police. Fundamentally, for Reicher, at the core of the dynamic between the police and the crowd is the formation of collective perceptions and identities by participants. In suggesting this, he reiterates the contention that the possibility exists for the escalation of force as the responses and counter-responses of the police and protesters become adjusted to one another.[42]

Given this sort of assessment, indiscriminate actions of the police in operations such as clearing the street become more understandable, though not necessarily condonable. The points made by Reicher and others suggest that reducing confrontation, at least in part, depends on reducing those things that reaffirm collective identifications of 'us' versus 'them' in events. Here, the use of tear gas against a large ensemble of individuals, as in Seattle, would generally reinforce such group identifications. Also, claims made by the police and government agencies just before the event about the imminence of serious violence against the police (for instance, 'Molotov cocktails'), later said to be baseless, probably did little to de-escalate confrontations.[43]

As the assessment of what happened and why at Seattle varied, so did the lessons learnt. The report of the Seattle Police Department made a series of recommendations about the policing of civil disturbances in the future, such as the need for greater resources, more elaborate planning and contingency response, better command systems, and the further provision of personal protective equipment. The Department further recommended expanding the training of officers in the use of chemical agents as this was seen as resulting in a low risk of injury to law violators and officers. The basic orientation of the recommendations was to more robustly implement the tactics and technologies deployed in the WTO conference.

At least one manufacturer shares the upbeat assessment of the appropriateness of the non-lethals employed in Seattle. Promotional material for JAYCOR includes a photo of a scene of police in all-black clothing and protective equipment standing over a group of individuals huddled together on the ground amidst clouds of chemicals. The caption reads: 'Jaycor's PepperBall™ Systems were successfully deployed by the Seattle Police Department at the World Trade Organization riots in December 1999. The widespread media attention in Seattle gave PepperBall™ national and international exposure.'[44]

The interpretation of Seattle as a failure to control rioters is predictably not shared among those groups critical of policing actions. Instead, the events represent a failure by responsible agencies to professionally enforce the rights of all parties to the conference. The reasons for this failure vary, sometimes stemming from concerns beyond the WTO conference itself. The Seattle National Lawyers Guild made sense of the events in the following way:

> It is the theory of this report [into the WTO protests], that because of the unchecked growth of the military industrial complex, especially in the area of law enforcement, and because of the undemocratic nature of the WTO and its policies, the response of the police to thousands of successful demonstrators became almost inevitable.[45]

In a similar fashion, CLGA interpreted the events as part of a wider clampdown on political dissent in the United States and elsewhere facilitated by the 'militarization of police practices'. To what extent the requirements between the military and the police are collapsing is thus a major concern for some, and part and parcel of how the appropriateness of non-lethal weapons ought to be understood. Along these lines, Morales[46] cites the US Army's Civil Disturbance Plans that refer to the need to draw on tactics and plans from operations other than war (as in peacekeeping in Bosnia) for the controlling of protests such as that in Seattle. This fear of the future incorporation of non-lethals into the policing of demonstrations is confirmed by a joint Russian-American conference that established among the desired functions of non-lethals the ability to 'neutralize mass demonstrations via dispersion and blocking of crowds and stopping means of transport'.[47] The underlying logic portrayed for critics is one of the further and further limiting of political dissent, in part facilitated by the introduction of new weapons. Such a 'chilling effect' has been denied elsewhere. As John Alexander maintains, '… dissent seems to be alive and well in American society. Non-lethal weapons have not changed that nor would their future availability make any difference in the amount of dissent or how it is displayed.'[48]

In examining the case of Seattle and public-order policing more generally, my goal has been to illustrate the multiple ways in which 'context' is drawn upon to make sense of situations. When questions are raised about the acceptability and legitimacy of the use of non-lethal weapons, the context at stake becomes a source of debate. Are police officers really responsible for using non-lethals in proscribed ways 'given' the volatility of contexts? How are identities of participants dependent on the context in question? What context should be drawn on for understanding the reasons for the recourse to non-lethals? The relevance of each of these aspects is, in turn, a matter of debate. It is in the unfolding of debates that the meanings attached to non-lethals become defined.

So far in this chapter, a number of possibilities have been offered about how the acceptability and effects of non-lethals vary by context: the context of use changes over time; the availability of some weapons changes the appropriate response to 'a' context; the use of certain technologies helps define the identity of individuals in that setting; the likelihood of adhering to rules depends on the context and the drawing on multiple contexts to make sense of the reasons for particular events. Each of these facets is an area where there is potential for disagreement. Any determinations of what happened in an event and why require making assumptions about the adequacy of evidence and presumptions about the interpretation of such evidence, where these issues are in turn bound up with assessments of the credibility and reliability of claims. The next section takes this further by focusing on how technology and its context of use are defined in relation to one another.

DIVIDING RESPONSIBILITY IN A DIVIDED LAND

Beyond public protests or riots in Western Europe and North America, non-lethal weapons have been deployed in areas of long-running conflict. Israeli security forces deploy a variety of non-lethal weapons in policing Palestinians, such as tear gas, plastic bullets and rubber ammunition. The latter come in two forms, rubber-coated steel bullets for aiming at groups and rubber-coated cylinders for targeting individuals.

136

A report in 1998 by B'Tselem, the Israeli Information Center for Human Rights in the Occupied Territories, into the use of rubber bullets and cylinders provides an illustration of how concerns about the adherence to rules, the contexts of the use of non-lethals, the characteristics of the technology, and notions of responsibility are defined in relation to one another.[49]

Rubber-coated ammunition fired from rifle-mounted canisters were introduced into Israeli security forces in the late 1980s as a non-lethal means of responding to riots and clashes between Palestinians and security forces. As with other kinetic-energy weapons, guidelines were produced to ensure that they caused minimal injury. In this case, the ammunition was not supposed to be fired under a distance of 40 meters and was only supposed to be used after other methods to disperse crowds had failed (for instance, tear gas and warning shots with live ammunition), the bullets were not supposed to be fired at children, and the cylinders were to be targeted at the lower portion of the body. While the number of injuries associated with these weapons has been difficult to calculate, B'Tselem estimated that at least 28 children and 30 adults were killed between 1988 and 1998.

The reasons for these deaths are contested. B'Tselem contended that both forms of rubber ammunition were best thought of as lethal weapons. Expert forensic opinion by Robert Kirscherner was drawn on which concluded that even when used in accordance with the firing guidelines such ammunition could cause death, particularly for certain susceptible persons such as children, the elderly, or those with heart conditions.[50] The high number of serious injuries and deaths caused by rubber bullets were taken by Kirscherner to imply that the guidelines for the use of this technology were frequently violated. Members of the Israeli Defense Forces and the Office of the Military Advocate were presented as giving a number of justifications and excuses for the deaths that had occurred: the difficulty of estimating firing distances, the unpredictability of conflict situations which meant that children or others entered the line of fire, or the aiming errors of soldiers in volatile circumstances. In other words, when the operational guidelines were 'violated', this was not deliberate. The deaths and injuries were the unfortunate consequence of a combination of attenuating situational and operational factors. As such, little disciplinary action or prosecutions against soldiers has been justified, expect for certain rare cases where officers flagrantly violated the guidelines. B'Tselem countered such justifications by arguing that the ammunition was far too inaccurate to be used at the distances suggested, that the guidelines were impractical, and that there had been obvious cases of the deliberate violations of the guidelines that had gone unpunished due to lenient prosecution practices. Instead of unintended consequences of volatile situations, the death and injuries represented the outcome of decisions where little significance was placed on Palestinian lives.

Debates about the merits of rubber-coated ammunition illustrate the range of social and technical issues at stake in discussions about the merits of particular non-lethal weapons. Determinations of where blame lies, if any exists, are the outcomes of assessments about the feasibility of altering the tactics, training, technologies and use-of-force accountability procedures. The debate is not merely one of saying whether the rubber ammunitions are actually best thought of as lethal or non-lethal weapons, but rather it is about how a weapon can practically be attributed with certain properties in conditions of conflict. Just what the context is for understanding the effects of rubber-coated ammunition is part of the debate at stake. So, are the important contextual features those immediate situational ones at the time of firing, or those to do with wider concerns about impunity for violations of the guidelines? Furthermore, determinations of what the characteristics of the weapon are, say its accuracy, are defined in relation to an understanding of the context at stake. Whether the deaths inflicted have been attributed to the inherent inaccuracy of the ammunition (and hence preventable through adopting alternative options) or to the outcome of the situational complexities (and hence inevitable and not the fault of the technology) is under dispute. The potential for the displacement of blame is considerable. It should be clear that this contestation is part of the process of constituting an understanding of the 'effects of technology'.

The definition of the context for consideration leads to further questions about the likely appropriateness of non-lethal weapons. After an intensification of hostilities between Israelis and Palestinians in September 2000, Amnesty International visited Israel and the Occupied Territories.[51] Between late September and mid-October 2000, Israeli security forces reportedly killed 100 Palestinians, including 27 children, during demonstrations. The delegates from Amnesty concluded that these deaths resulted from policing tactics and technologies that were more analogous to military methods of war-fighting than to law-enforcement methods based on the imperative to protect life. Demonstrations followed a predictable pattern. Protests culminated at a highly symbolic location where confrontations were sought by *both* demonstrators *and* security forces. Tensions would escalate as stones or sometimes petrol bombs were thrown. Rather than attempting to disperse crowds through non-lethal means such as tear gas, fairly well protected security forces resorted to live and rubber-coated ammunition within minutes. Amnesty argued that even if those situations had required the protection of life, other options (such as tear gas or warning shots) could and should have been employed. Through investigations of conflict scenes, members of the Amnesty mission concluded that there was a lack of control in even a random firing of ammunition. Frequent violations of security forces guidelines meant that half of the deaths during the time of the visit resulted from 'non-lethal', rubber-coated rounds.

The deaths were not simply seen as the result of inherent effects of the live and 'non-lethal' ammunition, but on some occasions resulted from Israeli and Palestinian security forces obstructing medical care provisions to those injured. Restrictions placed upon the movement in and out of a confrontation area, as well as the targeting of medical crews, meant that international standards for law enforcement to ensure medical assistance at the earliest possible moment were not followed. Noting the importance of such access illustrates the potential for a wide range of contextual issues to be brought to bear in assessing lethality.

To illustrate something of the negotiation about the merits of non-lethal weapons, in this case Amnesty International called on Israeli and Palestinian forces to adopt a range of weapons and ammunition that would allow for the differentiated application of force, as well as further self-defensive equipment to prevent the likelihood of having to resort to force at all. In this regard, the organization shares the sentiments expressed by Steven Aftergood in Chapter 4 (see p. 84) about the desirability of some non-lethal options. Yet, as various forms of non-lethal weapon and other options have been around for some time, the issues at stake are more complicated than the mere availability of weapons. As Amnesty argued, '[t]he use of lethal force involving considerable shooting, injury and death, rather than the use of non-lethal force, was clearly a considered choice'.[52] This calls into doubt the analysis offered in Chapter 4 (see p. 69) by the US Joint Armed Forces that the problem with the Israelis' use of non-lethal weapons was that they were technically inadequate and as such required the resort to deadly force.

'WE DON'T TAKE PLASTIC'[53]

The cases above illustrate the scope for the displacement and negotiation of responsibility when things are seen to go wrong. Various reasons are offered as to why deaths and injuries have taken place with certain weapons, thereby specifying those issues in need of redress. In some framings, the weapons themselves have come under scrutiny; this in the sense of their inaccuracy or lethality, regardless of how they are deployed. As a way of minimizing unfortunate outcomes and disputes about the use of force, one possibility might be the introduction of alterations or alternatives to existing non-lethals. If other types of kinetic-energy weapons were introduced into policing in Israel, for instance, perhaps this would enhance the legitimacy of force. Considering the potential of such strategy raises key questions: In what way are particular weapons seen as unacceptable and what arguments are offered for such assessments? How are contextual variations of technology handled in such evaluations?

As mentioned in previous chapters, the use of 'plastic bullets' or 'baton rounds' in Northern Ireland by the former Royal Ulster Constabulary and the British Army has been the source of significant debate since the 1970s. Numerous deaths, maimings, fractures, eye loss and other injuries have been attributed to the bullets. Throughout that time, security agencies have maintained that this option was (and is) necessary to ensure that rioters armed with petrol bombs and other such missiles could be kept at a safe distance from soldiers and officers. Without such an option, greater force would be required in close-quarter situations.[54]

Such an assessment has been countered elsewhere by groups such as the Committee on the Administration of Justice (CAJ) and the United Campaign Against Plastic Bullets. Opposition has stemmed from such reasons as the bullets' lethality, the lack of controls on their deployment, and the impunity surrounding their misuse.[55] With regard to the lack of controls, this includes the systematic firing of plastic bullets close to individuals and the targeting of vulnerable parts of the body. Far from being weapons against dangerous rioters, they have caused serious injuries to innocent individuals, including children. It was not until 1997 that the guidelines for the use of the bullets became public.

The lack of accountability before the law is illustrated in CAJ's account of the death of Nora McCabe.[56] In 1981, she was struck in the back of the head by a plastic bullet. The RUC first denied that the Land Rover alleged to be the source of the shot had fired anything at the location. By chance, though, a television crew had captured the actions of the patrol car at the spot of firing. The video evidence indicated that a puff of smoke emerged from the Land Rover at the scene of Nora McCabe's death despite no signs of conflict. An inquest inquiry found Nora McCabe innocent of rioting. Despite the contradiction of the video imagery with the accounts given by RUC testimony, no officers were prosecuted for this killing, or for perjury, afterwards. Indeed, no convictions have been given for any deaths or serious injuries from plastic bullets in Northern Ireland. Concerns have also been raised about the disproportional firing of these devices against Catholics, where this reflected the biases of the predominately Protestant background of RUC officers.

Proponents of the plastic round maintain that previous concerns about their safety and deployment no longer hold. A chief superintendent of the RUC, Colin Burrows, argues that in recognition of the problems associated with earlier versions of this round and its launcher, various innovations have been made to both.[57] These changes, coupled with more restrictive guidelines introduced in 1999, have greatly reduced injuries and deaths. While before 1994 a proportion of 1 in every 6,500 rounds fired resulted in a fatality, since that time, 13,500 have been fired with no deaths. With the adoption of new guidelines, the number of plastic bullets fired has dropped to an average of fewer than 100 per year in contrast to thousands per year before that. As part of the peace process, attempts are now under way to create a more religiously mixed police force in Northern Ireland.

Even with the significant decrease in the firing of plastic bullets in recent years, the lack of fatalities, and the wider reforms in policing now under way, opposition remains. In 2001, the chief commissioner of the Northern Ireland Human Rights Commission, a group set up to review the adequacy and effectiveness of human rights protection, called for an end to the bullets in riots.[58] Arguably, no amount of modification is likely to been seen as adequate for detractors. The bullets per se remain 'vicious killers'.[59] What is the basis of such a position and what might this suggest about the legitimacy of non-lethal weapons?

To a large part, much of the opposition still stems from perceptions of the lethality of plastic bullets. In the past, those opposed have drawn on the medical studies cited in the third section of the last chapter as a counter to claims made by successive governments about the relative safety of plastic bullets. As contended by individuals such as Burrows and as suggested previously, this lumping together of all types of plastic bullets and launchers used in Northern Ireland conflates the hazards associated with particular modifications. Herein, each version of the weapon should be assessed on a case-by-case basis. For critics, though, much more is at work in declarations of the unacceptability of such weapons than a reading of risk assessments or official casualty numbers. Debates about the merits of modified plastic bullets at once concern the effects of such weapons (as determined by tests and limited field experience) as well as wider issues about their symbolic importance. So despite claims about the safety and effectiveness of plastic bullets, they were not used on mainland Britain until 2002 and this not in crowd-control situations. Yet arguably, public-order disturbances there, such as the urban riots of 2001, were as intense as many experienced in Northern Ireland in recent years. The disparity in treatment is taken to reflect different standards for the acceptability of possible harm to different populations.

The sort of case-by-case approach given in statements about the safety of plastic bullets also takes for granted that those controlling and operating the weapons are trustworthy and reliable, and that they can be expected to act in prescribed ways. Yet, as suggested in the criticism above, the safety of weapons should not be divorced from past experiences regarding the impunity of users and governments from accountability. Take the 2001 introduction of the new L21A1 round as a case in point. After a £1.6 million ($2.4 million) research program that started in 1997, this round was offered as a latest refinement in the development of the plastic bullet and one deemed to be more acceptable. As mentioned in the third section of the last chapter, while expert government medical advice suggested that the bullet would pose greater dangers to the head when it was struck, the new and improved sights on the launcher were said to make the L21A1 a safer option. That is, so long as operational guidelines were followed. Yet, the argument of groups such as CAJ is that assumptions about the adherence to guidelines are naive. The extent of this is conditional. A return to previous levels of sectarian conflict in Northern Ireland might bring back past practices by security forces. Even from a mere feasibility standpoint, policing bodies have questioned the ability of officers to judge firing distances.[60] The failure to publish the full safety assessment of this round and any data on its ricochet potential means that those concerned about plastic bullets must trust decision-makers to act appropriately, with due regard for health concerns. Past experiences for critics are taken as reasons for doubting whether this is a wise course of action.

Presented in this way, debates about the merits of the plastic bullets share many of the same characteristics of other scientific and technical controversies, such as the safety of genetically modified organisms, mobile phones, or transport. Social scientists that have examined the underlying reasons for the public acceptance of science argue that attitudes often depend on the trust – based on experience and perception – held in the relevant institutions. Trust grows as institutions are seen as sufficiently policed, accountable, and open to incorporating the knowledge and concerns of various groups.[61] To simply refer to plastic bullets in Northern Ireland as an 'emotive' issue[62] for affected individuals and families, then, is arguably not helpful.

Furthermore, a reluctance to agree with official statements about the safety or hazards of particular initiatives can stem from orientations to underlying assumptions and commitments behind particular claims. As Wynne argues, for instance, opposition to animal growth-hormones in Europe derived from what were held to be unrealistic assumptions about the feasibility of strictly controlling the administration of such biochemicals in real-life circumstances.[63] In other words, assessments of

142

risk vary in alternative models of the contexts in which technologies are supposed to function. In relation to criticisms about the lack of accountability of plastic-bullet use in Northern Ireland, for instance, starting in 2001, the Northern Ireland Police Ombudsman was supposed to be immediately informed about the discharge of plastic bullets so that, if deemed prudent, an investigation could be conducted of the acceptability of the use of force and its adherence to the guidelines. However positive such measures, the Human Rights Commission argues that is it doubtful that the Office of the Ombudsman will be in a position to be able to conduct thorough reviews should a high number of bullets be fired over a short time period.[64]

ADDITIONAL REMARKS

The previous section indicated something of the assumptions and commitments that underlie contrasting assessments of the merits of non-lethal weapons. As was argued there and in the preceding sections, determinations of the acceptability of non-lethal weapons (statements about what they are for and their effects) are defined in relation to a context. 'Context' itself is an elastic resource, which can refer to a variety of background assumptions about the drivers of actions or situational factors that hinder or encourage certain forms of behavior.

Disagreements about events are bound up with the mutual definition of context and technology. Whether one believes the official justification that riots in the West Bank are just so unpredictable that some deaths are unavoidable with non-lethal rubber-coated ammunition is a key factor in whether one downplays or decries the deaths attributed to these bullets. Technical predictive assessments of the characteristics of such weapons, such as their lethality, rely on assumptions about how they are used in practice (such as whether they are fired from a certain distance). Determinations of lethality also inform decisions about whether deaths resulted from unavoidable confusion or unacceptable expectations. So, as mentioned above, the high level of deaths to Palestinians because of 'non-lethal' ammunition was taken as an indicator of the inappropriateness of depending on the weapons being fired appropriately.

To say that there are multiple contexts at work in evaluations and different ways of making sense of context and technology is not to suggest that all accounts are equally valid, or that debates about the acceptability of the recourse to non-lethal weapons are not resolvable for practical purposes. Various claims and counterclaims have been offered in this chapter in relation to specific debates. The reader has no doubt found some evaluations more compelling than others. In making such evaluations, information outside this text was probably drawn on as well as assessments of the motives, identity and credibility of those making claims (including the author). While some may have no problem believing that Kenyan security forces have acted in malicious ways, the suggestion that officers did so in Northern Ireland might not attract the same support. Such determinations about the cause for effects or events do not result from a straightforward description of technology.[65] Rather than just asking what the effects of weapons really are, as if this could be stated once and for all without scope for debate, this analysis has sought to see how certain devices have certain characteristics attributed to them.

Just as the accounts above have drawn on or implied a certain context for making sense of the likely operation of technology, so, too, has my analysis. Just as I have presented debates about non-lethals revolving around appeals to what is feasible in real-life situations, so my analysis has rested on appeals to what is likely in the 'real world'. In this regard, it has been assumed and suggested that the strict adherence to rules in the use of force is highly unrealistic. This supposition has derived from my understanding of criminological, sociological and historical studies. The extent to which such general findings are deemed relevant to the types of issue at hand in this book depends, at least in part, on a consideration of the credibility of such studies and their applicability to situations discussed.

Later chapters further elaborate many of the themes and issues raised in this one. Chapter 8, for instance, examines concerns about discretion, rule adherence and responsibility in relation to a specific case: the introduction of CS sprays in Britain. The next chapter, though, further considers the negotiations of context and technology in attempts to control and regulate non-lethal weapons, as initially discussed here in relation to plastic bullets in Northern Ireland. While discussions about the contested status of technology and its context might seem abstract at times, such issues are highly pertinent and significant in relation to proposals about whether controls might need to be placed on non-lethal weapons.

NOTES

1. Someku, I., 'Death Toll Passes 130', *The Independent*, 11 May 2001.
2. This event followed a similar one in Zimbabwe, see Blair, D., '12 Crushed in World Cup Qualifier', *The Telegraph*, 10 July 2000.
3. Amnesty International, *Stopping the Torture Trade* (London: Amnesty International, International Secretariat, 1997).

4. Amnesty International, *Annual Report 1998* (London: Amnesty International Secretariat, 1998).
5. Hunt, J., 'Police Accounts of Normal Force', in V. Kappeler (ed.), *The Police and Society*, 2nd edn (Prospect Heights, IL: Waveland Press, 1999). See as well Bayley, D. and Garofalo, J., 'The Management of Violence by Police Patrol Officers', *Criminology*, 27, 1 (1989), 1–27; Waddington, P. A. J., *Policing Citizens* (London: UCL Press, 1999); and Jefferson, T., *The Case Against Paramilitary Policing* (Milton Keynes: Open University Press, 1990).
6. For instance, the term 'spray and pray' refers to the panic practice of officers firing multiple and poorly aimed shots in an attempt to hit a target rather than shooting with control; see 'Preventing 'Spray and Pray''', *Law and Order*, September 1999.
7. Hunt, 'Police Accounts of Normal Force', 321.
8. Adams, K., 'Measuring the Prevalence of Police Abuse of Force', in W. Geller and H. Touch (eds), *And Justice For All* (Washington, DC: Police Executive Research Forum, 1995), 61–97.
9. See Maguire, M., Morgan, R. and Reiner, R. (eds), *The Oxford Handbook of Criminology* (Oxford: Clarendon Press, 1997).
10. Bittner, E., *The Functions of the Police in Modern Society* (Cambridge, MA: Oelgeschlanger, Gunn & Hain, 1970); Ericson, R., *Making Crime* (Toronto: Butterworth, 1981); and Ericson, R., *Reproducing Order* (Toronto: University of Toronto Press, 1982).
11. For an elaboration of such issues see e.g. Holdaway, S., 'Discovering Structure: Studies of the British Police Occupational Culture', in M. Weatheritt (ed.), *Police Research* (Aldershot: Avebury, 1979).
12. See Goldstein, J., 'Police Discretion Not to Invoke the Criminal Process: Low Visibility Decisions in the Administration of Justice', *Yale Law Journal*, 69 (1960), 543–94.
13. For instance, see McEwen, T., 'Policing on Less-than-Lethal Force in Law Enforcement Agencies', *Policing*, 20, 1 (1997), 39–59.
14. In the United States, this is stated in Supreme Court decisions such as TENNESSEE v. GARNER, 471 US 1 (1985) and GRAHAM v. CONNOR, 490 US 386(1989).
15. Waddington, D., 'Key issues and controversies', in C. Critcher and D. Waddington (eds), *Policing Public Order* (Aldershot: Avebury, 1996), 166.
16. Ericson, R. and Doyle, A., 'Globalization and the Policing of Protest', *British Journal of Sociology*, 50, 4 (1999), 589–608.
17. Lumb, R., and Friday, P, 'Impact of Pepper Spray Availability on Police Officer Use-of-Force Decisions', *Policing*, 20, 1 (1997), 136–48.
18. See also Kock, E., Kemp, T. and Rix, B., *Assessing the Expandable Side-Handled Baton* (London: Police Research Group, 1993).
19. Wynne, B., 'Unruly Technology: Practical Rules, Impractical Discourses and Public Understanding', *Social Studies of Science*, 18, (1998), 147–67.
20. Quoted from Robinson, J. 'Disabling Chemical Weapons', Presentation to PUGWASH Study Group on Implementation of the CBW Conventions, 27–29 May (Den Haag: PUGWASH, 1994).
21. Dando, M., *A New Form of Warfare* (London: Brassey's, 1996), Chapter 5.
22. See Shalom, S., 'Bullets, Gas, and the Bomb', Z (1991). See www.zmag.org/zmag/articles/ShalomWeapons.html
23. Omega Foundation, *Crowd Control Technologies* (Luxembourg: European Parliament, 2000), Appendix 6.
24. For a further discussion see Feenberg, A., *Questioning Technology* (London: Routledge, 1999) and Winner, L., *Autonomous Technology* (Cambridge, MA: MIT Press, 1997).
25. Bell, F. and Thomas, B., 'Police Use of CS Spray', *Mental Health Care*, 1, 12 (1998), 404.
26. Ibid.
27. Kraska, P. and Kappeler, V., 'Militarizing America Police', in Kappeler (ed.), *The Police and Society*, 463–79.
28. Weber, D., *Warrior Cops*, CATO Institute Briefing Paper 50 (Washington, DC: CATO, 1999).
29. For similar arguments about Britain see Northam, G., *Shooting in the Dark* (London: Faber & Faber, 1988) and Jefferson, *The Case Against Paramilitary Policing*.

30. For a report commissioned by Seattle Mayor Schell see McCarthy, R. & Associates, *An Independent Review of the World Trade Organization Conference Disruptions in Seattle, Washington November 29–December 3, 1999* (Seattle: McCarthy, R. & Associates, 2000).
31. See e.g. Seattle National Lawyers Guild, *Waging War on Dissent* (Seattle: Seattle National Lawyers Guild, 2000); Amnesty International, *Amnesty International Calls for an Inquiry into Police Actions at WTO Talks in Seattle* (London: Amnesty International, 1999).
32. Allen, T., 'Chemical Cops', *In These Times*, 3 April (2000).
33. Seattle National Lawyers Guild, 'Seattle as a Model of A New Type of War', in *Waging War on Dissent* (Seattle: Seattle National Lawyers Guild, 2000), 13–22
34. Seattle Police Department, *After Action Report – World Trade Organization Ministerial Conference* (Seattle: Seattle Police Department, 2001).
35. Seattle City Council, *Report of the WTO Accountability Review*, 14 September (Seattle: Seattle City Council, 2000).
36. Ibid., 2.
37. Ibid., 3.
38. Ibid., 12.
39. The Committee for Local Government Accountability, *Citizen Group's Response to Seattle's Civil Emergency and WTO Accountability Review*, 14 September (Seattle: The Committee for Local Government Accountability, 2000).
40. See Stoot, C. and Reicher, S., 'Crowd Action as Intergroup Process', *European Journal of Social Psychology*, 28 (1998), 509-529 and Reicher, S. "The Battle of Westminster" *European Journal of Social Psychology*, 26 (1996), 120–46.
41. And indeed as part of the ascription of blame in public-order situations it has been suggested that the police often are not interested in making such a distinction but instead see the world through various stereotypes; see Saunders, M., 'A Preventive Approach to Public Order', in Critcher and Waddington (eds), *Policing Public Order*.
42. Della Porta, D., *Social Movements, Political Violence, and the State* (Cambridge: Cambridge University Press, 1995) and Della Porta, D. and Reiter, H. (eds), *Policing Protest* (London: University of Minnesota Press, 1998).
43. See The Committee for Local Government Accountability, *Citizen Group's Response to Seattle's Civil Emergency and WTO Accountability Review* and Ream, T., 'False Police Reports are Part of Police Strategy', in *Waging War on Dissent* (Seattle: Seattle National Lawyers Guild, 2000).
44. Jaymark, *2000 Capabilities Brochure* (San Diego, CA: Jaymark, 2000).
45. Richmond, P., 'Seattle as a Model of A New Type of War', in *Waging War on Dissent* (Seattle: Seattle National Lawyers Guild, 2000), 4.
46. Morales, F., 'Non-lethal Weapons: Welcome to the Free World', *Covert Action*, 70, April–June 2001, 6–15.
47. Selivanov, V., Klochikhin, V. and Pirumov, V., 'Modern Views on the Development and Application of NLW in Anti-Terrorist and Peacekeeping Operations', in *Non-lethal Weapons*, 17–3.
48. Alexander, J., *Future War* (New York: St Martin's Press, 1999), 189.
49. B'Tselem, *Death Foretold*, December (Jerusalem: B'Tselem, 1998).
50. Kirscherner, Robert, 'Forensic Aspects of Rubber Bullet Injuries', in B'Tselem, *Death Foretold*, December (Jerusalem: B'Tselem, 1998).
51. Amnesty International, *Israel and the Occupied Territories Excessive Use of Lethal Force*, 19 October (London: Amnesty International, 2000).
52. Lethal force in this case is meant to include the use of rubber-coated bullets because of their dangers and their persistent use in violation of the guidelines.
53. I have been told that his phrase has been a protest shout at some nationalist demonstrations in Northern Ireland to the firing of plastic bullets.
54. For a more recent statement of this argument see Her Majesty's Inspector of Constabulary, *Primary Inspection, Royal Ulster Constabulary* (London: HMSO, 1996).
55. United Campaign Against Plastic Bullets, *5 Reasons to Ban Plastic Bullets* (Belfast: United Campaign Against Plastic Bullets, 1997).
56. Committee on the Administration of Justice, *Plastic Bullets: A Briefing Paper* (Belfast, 1998), 4.

57. Burrows, C., 'Operationalizing Non-lethality', *Medicine, Conflict, and Survival*, 17, 3 (2001), 260–76.
58. Cowan, R., 'RUC Rejects New Call to Ban Plastic Bullets', *The Guardian*, 19 July 2001.
59. United Campaign Against Plastic Bullets, *5 Reasons to Ban Plastic Bullets*.
60. Steering Group for Patten Report Recommendations 69 and 70 Relating to Public Order Equipment, *A Research Programme into Alternative Policing Approaches Towards the Management of Conflict*, 21.
61. Wynne, B., 'Public Understanding of Science', in S. Jasanoff et al. (eds), *Handbook of Science and Technology Studies* (Sage: New York, 1995) and Irwin, A. and Wynne, B. (eds), *Misunderstanding Science?* (Cambridge: Cambridge University Press, 1996).
62. Acres, P., 'Conflict Management Portfolio', Presentation at Safer Restraint Conference, London, 17 April 2002.
63. Wynne, B. 'Technology Assessment and Reflexive Social Learning' , in Rip, A., Thomas, M. and Schot, J. (eds), *Managing Technology in Society* (London: Pinter, 1995).
64. Northern Ireland Human Rights Commission, *The Recording of the Use of Plastic Bullets in Northern Ireland* (Belfast: NIHRC, 2001).
65. See Grint, K. and Woolgar, S., *The Machine at Work* (Cambridge: Polity, 1997), Chapter 6.

7

Controlling Evaluations:
The Prospects for Prohibitions

Throughout the previous chapters, competing versions of what non-lethal weapons are and what they do have been presented. Such interpretations are based on assumptions about, for instance, the manner in which devices are used in practice. Evaluations of their likely appropriateness depend on the motives and competencies attributed to users, the characteristics of the technology, and the functioning of a 'weapon' in its 'surroundings'. Where the salience of such alternative accounts becomes stark is in determinations about controls and restrictions on non-lethals. Pressing questions arise about when it is justified to close down deliberations and cut through associations of individuals and technologies in order to take as fixed the characteristics of weapons or the likely actions of their users in order to support certain control measures. Various aspects of the formal regulation of non-lethal weapons are examined in this chapter: export controls, international humanitarian law and arms-control treaties. Much of the argument focuses on the military-related aspects of non-lethals where international standards are relevant. The argument presented in this chapter will set out bases for analysis that will be taken up in Part III.

CONTROLS FOR PROLIFERATION

To avoid the proliferation of weapons of mass destruction and conventional ones being sent to what are deemed undesirable destinations, most countries have enacted export-control systems. These include provisions requiring companies and sometimes government agencies to obtain licences for the export of weapons or 'dual-use' equipment that has security and civilian applications. International agreements also require states to take responsibility for the consequences of their actions. In the case of arms transfers, the lawfulness of transfer from a country would be called into doubt if the recipient were considered intent on committing what would be a wrongful act in the exporting country.[1] As innovation leaders and prominent weapon manufacturers, the United States and Europe have controls which are key for minimizing the proliferation of non-lethal weapons. This section surveys the control mechanisms in place with a view to asking what is known about the outcome of such measures and about how decisions are taken.

Regulatory oversight of the trade in arms in the United States and Europe consists of a number of mechanisms such as procedures and criteria for the granting of licences, reports on arms transfers, and parliamentary or legislative scrutiny of approved transfers. The terms of coverage, however, vary from country to country, as do the procedures for publicly reporting transfers. The adequacy of control measures has been called into question by various commentators who maintain that existing export controls are characterized by numerous deficiencies and loopholes that do little to prevent US and European governments and companies from contributing to human-rights violations.[2] In recent years, the sale of arms to Indonesia in support of its former illegal occupation of East Timor,[3] the supply of hundreds of millions of dollars' worth of arms to all the countries involved in the Congo War,[4] and the shipment of arms to Rwanda by British and French nationals during the genocide in 1994[5] have been some of the most condemned transfers.

While major conventional weapons (tanks, aircraft, artillery) and weapons of mass destruction are monitored by various international arrangements, much less is known about the international trade in small arms or security equipment, such as existing chemical, electrical and kinetic non-lethal weapons. Limitations in the transparency make it impossible to know the full extent of the trade in these devices. A report by the NGO Saferworld in 2000, for instance, argued that the quantity and quality of information provided by European governments varied widely.[6] Most of these reporting systems make it impossible to know the amount of arms exported in a given transfer, the specific types of equipment in such transfers, or their specific recipient. While, in recent years, many European governments have introduced human-rights clauses into their export-control procedures, that require factoring into account the human-rights record of the recipient country, little is known about how such criteria are applied in practice.

In this general situation, it is quite difficult to establish systematic information on the trade of non-lethal weapons or to determine how decisions are made about exports. In the past, the appropriateness of certain transfers has been called into doubt. During 1997 in Zambia, for instance, tear-gas canisters were reportedly used against peaceful political protesters. One such event was the break-up of a march in Lusaka city center by the Zambian police with canisters manufactured in

149

Britain. Despite complaints about such practices, the British government subsequently granted further licences for the export of tear-gas grenades and irritant ammunition to Zambia.[7] After reports about the use of British-manufactured tear gas in human-rights violations in Kenya mentioned in the last chapter, the UK government later refused a licence for the sale of £1.5 million ($2.25 million) worth of riot-control equipment. Deprived of their British source, the Kenyan authorities turned to the French manufacturer, Nobel Sécurité. Reports of tear gas being employed by the authorities against peaceful demonstrators in enclosed spaces were subsequently made.[8] This later French transfer took place despite the establishment of the European Union Code of Conduct on Arms Exports in 1998 that was supposed to provide minimum standards for arms exports across the union. However, no agreed-upon list of security and police equipment to be covered by the Code has been established at the time of writing.

Long before the round of conflict that engulfs the area during the writing of this book, the legitimacy of tear-gas use by Israeli security forces in Israel and the Occupied Territories has been questioned. While tear gas might provide effective and relatively benign use-of-force options if used appropriately, a key concern is whether this has indeed been the case. Amnesty International have claimed that up to 50 people died between 1987 and 1993 because of the release of tear gas in enclosed spaces: an assertion disputed by Israeli authorities.[9] When the outbreak of widespread violence began again in late 2000, France and Germany initiated undeclared embargoes of defense equipment to Israel, including tear gas.[10] The grounds for these policies were not articulated at the time. It seems reasonable to assume that even non-lethal weapons were included in such bans because of the possibility for unacceptable consequences. Israeli officials reportedly charged the French government with 'doublespeak' for condemning the resort to live ammunition in controlling riots, but then failing to provide non-lethal weapons to replace them.[11] Criticisms in Chapter 6 suggest that the use of live ammunition and non-lethal forms are not distinct approaches to force in handling disturbances. To the extent that the decisions to ban the sale of tear gas rested on such considerations, they reiterate the themes discussed in that chapter about whether one takes as a frame of evaluation a case-by-case focus for decisions where a possibility is ever present for beneficial use, or whether one instead takes into account perceptions of past experience.[12]

The importance of such alternative approaches will continue to be a key concern. Chapter 2 discussed debates about the merits of helicopter gunships in attacks in the West Bank and Gaza strip. These weapons were said to allow for the pinpoint targeting of key strategic facilities that minimized potential casualties. Drawing on internal Israeli Air Force documentation, the journal *Defense News* claimed that largely US-manufactured Apache and Cobra helicopters with missiles – originally intended to support ground forces in military maneuvers – had been deployed on more than fifty assassination missions during 2001. The stated rationale of such missiles was to avoid collateral damage to surrounding areas, though their use has drawn widespread criticism as excessive force or extra-judicial killings. In the search for more appropriate use-of-force options in urban terrains, an Israeli Air Force base commander is reported to have said:

> We need to start talking about weaponry that can disable, but not kill. Just like the ground forces have rubber bullets and tear gas, we need to develop for helicopters a type of weapon that doesn't kill. This is a different type of war, and we don't always need to kill to achieve objectives.[13]

Stated in terms of the desire to reduce death in conflict, non-lethals might be regarded as more humane and desirable options in a given situation than conventional missiles. Yet, whether such weapons could or would function in such a proscribed manner or whether their use would be limited to situations in which lethal force would otherwise be necessary and justified are matters for debate. There would certainly seem to be a market for non-lethal missile technologies in Israel; a key question is, though, are there any sellers?

While much of the past interest in the development of non-lethal weapons has been concentrated in Western Europe and North America, production capabilities are now being globalized. In terms of proliferation, with this comes the danger that existing controls might be undermined or safety standards might be compromised. Lewer and Feakin, for instance, document how India has moved from being reliant on the United States, Britain and France for chemical riot-control weapons to being a manufacturer and exporter of such devices.[14] Information about exports from India is strictly classified, though. The range of munitions produced includes weapons utilizing CR,[15] sometimes in combination with CS or CN. As mentioned in Chapter 3, despite its potency, CR has not been taken up in the United States and Europe because of its health hazards and unknown long-term effects. Weapons that combine CR with CS or CN pose major decontamination problems as well because CR retains its potency in water. Those attempting to relieve irritation from exposures with water would end up worsening the extent of contamination.

As part of this overall globalized production and distribution of defense and security equipment, other possibilities of circumventing existing controls have emerged. In many European countries, for instance, manufacturers can avoid international and national regulations by arranging for their arms and equipment to be produced by a foreign non-EU company through licenced production deals. In such arrangements, the non-EU company produces identical products to those manufactured in Europe. The difference is that this technology is only bound by export controls of the foreign manufacturing country. Even when there are supposed to be rules in place preventing such an undermining effect, the controls for licenced production are unclear, vague, or not implemented.[16] Another way of getting around direct export licence controls is through brokering. Arms brokers can arrange the shipping and delivery of weapons and equipment so long as they do not physically enter the EU. Many of the transfers would not receive an export licence if the declared intention were to export the equipment directly from an EU country. In this way, even those countries that do not produce particular arms or equipment can participate in the trade of that technology. For instance, in 1996, a company representative from Belgian Business International claimed that the firm brokered electroshock technology to a country in central Africa. It avoided Belgian controls by using African and Latin American manufacturers as the suppliers.[17]

On the Torture Trail

Debates about the acceptability of particular transfers do not only turn on questions about the identity of using them. Of all the weapons that offer a potential for facilitating ill-treatment or torture, the category of electroshock equipment has come under particular scrutiny. Groups such as Human Rights Watch and Amnesty International have reported electroshock technology used in cases of torture and ill-treatment in 70 countries, both 'developed' and 'developing' nations including the United States, Turkey, China, Saudi Arabia and Egypt.[18] Testimonies from victims of the use of shock have been collected to highlight the potential of this technology for ill-treatment. One victim from the former Zaire said that:

> This time they worked on me again and again with the electric baton on the nape of the neck and in the genitals and it hurt so much that even now when I speak it is difficult to keep my head still as the back of my neck hurts very much ... This type of weapon ... I could really call it something really horrible – immoral – because those people who make it for

torture, they don't test it on their own bodies and they don't know the pain it causes. They do it to make other people suffer quite simply to make money. It's very sad.[19]

In the past, the United States, Germany and France were the main manufacturers of electroshock devices, though China and Taiwan have recently become prominent exporters. In 2001, the US government approved export applications for nearly $24 million worth of electroshock equipment for non-military use, though, because of a recent change of policy, it has stopped releasing information about the destinations of such exports.[20] Amnesty International has recommended stringent controls on the export and use of such equipment. Remotely activated electroshock restraint belts worn by prisoners during their transportation or by defendants in court hearings in the United States have come under particular criticism.[21] The organization has called for a ban on such devices altogether because they are cruel, inhumane and degrading; a verdict shared by the UN Committee Against Torture.[22] Whatever their health implications, the belts are said to induce unacceptable levels of fear and anxiety in wearers.

While some acknowledge the potential for abuse with this class of technology, such abuses are said not to justify bans or criticisms of the technology per se. As any object can act as an instrument of torture or ill-treatment, it is the intent of the users that is the problem. Non-lethal weapons in this regard are no different than cattle prods, ice picks, hammers or cigarettes.[23] The counter to such points is that the ease of using electroshock technology, the inability of users to gauge its effects, and the lack of residual body marks on the body if used in a particular manner make such weapons ideal for abuse. In the case of the Zairian above, Amnesty International reports 'he was first beaten with sticks, before an officer stopped the beating, saying "*it will leave scars and will get complaints from Amnesty International*".' The officer then ordered his men to use an electro-shock baton instead.'

It is not only human-rights organizations that have taken a stern critical position about electroshock weapons. For purposes of export controls from Britain, electrical devices such as taser and stun batons have been classified as instruments of torture since 1997 and their export is strictly forbidden. The British government has also publicly supported such a categorization for the European Union as a whole.[24] Some support even exists for such classifications in the United States.[25]

As with other weapons, often more is at stake in deliberations about controls than just technical or operational concerns. What might be labeled as 'symbolic dimensions' also come into play. In 2001–2, the UK government began evaluating taser electroshock weapons for use by the police. That was called into question by Amnesty International, who described it as providing a 'seal of approval' for the uptake of such equipment elsewhere.[26] The treatment of electroshock devices as both instruments of torture when it comes to export but effective internal law-enforcement options is arguably tension-ridden. Those countries wishing to uptake such technology could argue that what is good enough for Britain is good enough for them.

That such distinct ways of making sense of a class of technology exist is not surprising for some. Rejali contends that there are two competing stories about electrical weapons in public discussions. One story locates their origin and diffusion in the unreason and evil of particular torturers' practices.[27] Here, technologies and techniques supposedly invented by the Nazi Gestapo spread after the Second World War to the French in Algeria, the CIA in Vietnam, and the Argentinian police, eventually to become transformed into routine devices for policing worldwide. A second story for making sense of electric weaponry locates their origin in the steady and rational evolution of technological know-how to solve security problems. One such current application would be introduction of these devices into airline cockpits. While neither story is fully accurate for Rejali, they do enable individuals to both condemn electroshock technologies as instruments of torture, while also tolerating their deployment in other situations. He emphasizes that the background context given to this technology in deliberations is crucial in its legitimization or delegitimization.[28] This fight over origin can be seen in current evaluations of electroshock technologies. As part of a medical review of the safety of tasers for a British police force, an evaluation concluded that 'depending on how their introduction might be publicized in the media, their use might be construed as a potential weapon of torture'. The medical authors advised that 'the media portrayal of the introduction of these weapons needs to be handled very carefully'.[29]

INTERNATIONAL LAW AND TREATIES

Proliferation is not the only area where determinations about weapons enter into the regulation of technology. International law and arms-control treaties place various constraints on the means and methods of warfare and afford protection for persons or property that might be affected by a war but not actively partaking in it. The laws of war and arms treaties have long histories and have been topics of significant academic, public and policy debate.[30]

In this section, I want to survey the principles of international law and arms control as they relate to non-lethal weapons. In doing so, the aim is rather modest. Consideration is paid to what these legal provisions might mean for constraints to the development of certain weapons. Given the possibility that existing controls might place major limitations on the legality of certain technologies detailed in Chapter 3 (especially chemical and biological weapons – see pp.158–60 in this chapter), much deliberation is now under way regarding whether specific non-lethals, or their use in particular situations, comply with or violate the principles of war.[31] Debates about international law now routinely figure in conferences and publications about non-lethals. Here, I want to examine how characterizations and classifications of contested weapons are made in relation to international agreements. In this process there are basic issues about how claims are advanced and countered; how the effects and characteristics of weapons are defined, identified, and interpreted; and what issues are at stake in deliberations about technologies' legality.

The laws of war specify a number of general principles and prohibitions regarding the choice and deployment of weaponry. Figure 7.1 lists a number of the international declarations and conventions identified by Lewer and Schofield as pertinent to non-lethal weapons.[32] The major principles underlying these agreements relate to notions about discriminate attack, unnecessary suffering and superfluous injury, and *hors de combat*.[33] While any weapon might be used indiscriminately, the Geneva Conventions, Additional Protocol I Article 51(4) prohibits those attacks that are not directed at a specific military objective, as well as those attacks that employ a method or means which cannot be directed at a specific military target or otherwise limited. With regard to the second principle, as first specified under the 1907 Hague Regulations, military forces must refrain from causing superfluous injury or unnecessary suffering. Following on from this, *hors de combat* prohibits attacks against military personnel that no longer pose a threat. Fidler argues that in relation to non-lethals, this principle places both positive duties for aftercare, as well as negative duties to refrain from attacks once a target is incapacitated.[34] As expressed by these principles and in the agreements listed in Figure 7.1, the methods and means of conducting war are not unlimited.

Blinding Lasers

Just what such principles, agreements and limits to warfare ought to mean in practice, though, must be determined in practice. General rules and standards must be applied in specific cases where there is potential for debate over both the interpretation of technology and its context, and the meaning of the standard. As mentioned in Chapter 1, during

FIGURE 7.1
INTERNATIONAL CONVENTIONS AND DECLARATIONS RELEVANT TO NON-
LETHAL WEAPONS

- The Lieber Code – 1863

- Declaration of St Petersburg – 1868

- Hague Declaration (IV, 2) Concerning Asphyxiating Gases – 1899
  Hague Declaration (IV, 3) Concerning Expanding Bullets – 1899
  Hague Convention (IV) Respecting the Laws and Customs of War on Land – 1907

- The Protocol for the Prohibition of the Use in War of Asphyxiating, Poisonous or Other
  Gases, and of Bacteriological Methods of Warfare – 1925

- Geneva Convention – 1949

- Convention on the Prohibition of the Development, Production and Stockpiling of
  Bacteriological (Biological) and Toxin Weapons and on their Destruction – 1972

- Convention on the Prohibition of Military or Any Others Hostile Use of Environmental
  Modification Techniques – 1977

- Geneva Conventions, Additional Protocols I and II – 1977

- UN Convention on Certain Conventional Weapons – 1980

- Chemical Weapons Convention – 1993

1995, members of the UN Convention on Certain Conventional Weapons deliberated the establishment of controls on blinding-laser weapons. As already suggested, much of this debate turned on claims about technology and its context. Rather than taking the intent to kill in warfare as the standard against which blinding lasers should be compared, those pressing for prohibitions suggested that, given the effectiveness of lasers in practice, the standard for assessment ought to be the casualty rates incurred in warfare in general. These were lower than might be expected. On 13 October 1995, Protocol IV to the Inhumane Weapons Convention was adopted. Article 1 states:

> It is prohibited to employ lasers specifically designed, as their sole combat function or as one of their combat functions, to cause permanent blindness to unenhanced vision, that is to the naked eye or to the eye with corrective eyesight devices. The High Contracting Parties shall not transfer such weapons to any State or non-State entity.[35]

In the future, proposed laser systems will be judged against this criterion, at least in signatory countries.

Even where such agreements have been reached, though, the arguments of previous chapters would suggest that their meaning in practice will be a matter of negotiation. In the case of the blinding-laser protocol, for instance, various conditionalities are stipulated. Lasers are incorporated into a variety of military technologies, such as range finders and target designators that improve the discrimination capabilities of some weapons. These lasers might cause inadvertent blindness if their beam passes over eyes. The established role of lasers in advanced weaponry meant that a total ban on their use in warfare was unlikely. State Parties to the Convention on Certain Conventional Weapons sought to distinguish blinding as a method of warfare versus blinding resulting from incidental or collateral effects of the otherwise legitimate employment of lasers. While Protocol IV restricts the former, it does not restrict the latter.

With this differentiation comes the difficulty of classifying those weapons that are really meant to blind and those that do so inadvertently. In early 1995, a Chinese company exhibited a laser device (the ZM-87 Portable Laser Disturber) for international export that could damage the human eye or disrupt sensor equipment. Suggestions have been made that since the Protocol agreement, something remarkably similar to the ZM-87 has been incorporated into the defense system for a Chinese tank, here ostensibly to disable the guidance optics of weapons targeting the tank.[36] Such allegations raise a variety of concerns about how easily defensive measures might be able to be turned into personnel-blinding weapons. Many of the same questions have been asked about the export from the United States to Israel of a laser air-defense system.[37]

The threshold at which blinding begins is another area for classification disputes. Human Rights Watch contended that after the 1995 agreement, US military agencies modified the intensity of laser systems then under development from being potentially blinding to 'merely' dazzling. Where dazzling ends and blinding begins, though, is not clearcut. The effects of lasers vary by environmental conditions and how they are used in practice. Even assuming laser systems do not cause blindness (and therefore do not fall foul of Protocol IV), Human Rights Watch expressed concern that such developments might subvert the purpose and intent of the agreement. As it argued:

> The emergence of operational lasers intended to attack eyes ultimately has the effect of undermining the humanitarian and nonproliferation objectives of Protocol IV [Blinding

Lasers]. U.S. military personnel are thus more likely to face blinding lasers on the battlefield. The emergence of anti-personnel lasers in the civilian world also opens the way for criminal elements to obtain anti-personnel lasers. For the U.S. to research and deploy these systems undermines the 'norm' against blinding that Protocol IV represents.[38]

Just whether such a norm was set up by the Protocol and whether it is being undermined by dazzling lasers are not topics of unanimity. In the case of blinding lasers, then, it is possible to see something of the scope for disagreement about how distinctions are made between weapons, about the goals of particular treaties, and about how classifications made in treaties can be policed. These are recurring themes in attempts to control weapons.

Chemical and Biological Weapons

The case of non-lethal chemical weapons provides a further illustration of the possibility for negotiation about the scope of agreements and what is at stake in disputes. Dando provides an account of the history of international control of chemical agents such as CS and CN.[39] The 1925 Geneva Protocol provided substantial constraints on the use of asphyxiating, poisonous, or other gases that it stated had been 'justly condemned by the general opinion of the civilized world'. The ability to circumvent these constraints is illustrated by later actions of some governments. As Dando argues, while the United States did not become a signatory to the protocol by the time of the Vietnam War, it had been bound by its conditions under customary law. Despite initial agreements in the 1930s that agents such as tear gas and herbicides were included within the protocol,[40] the US government later interpreted that these chemicals were outside its scope. As officials argued during the Vietnam War, such weapons might provide a more humane option in certain settings.[41] It was not until the mid-1970s that the United States adopted the 1925 Protocol. This was done with various provisos enabling harassing agents such as CS to be used in defensive military actions. Similarly, in 1970, the British government deemed CS to be outside of the Protocol because it was a smoke of low lethality rather than a lethal gas. The then use of CS in Northern Ireland, for some commentators, meant that this government had a clear desire to find some pretense for refraining from using on one's citizens chemicals 'justly condemned by the general opinion of the civilized world'.[42]

That the United States should have chosen to employ CS in Vietnam was regrettable for Dando because of the continuing interest it drew to chemical weapons as a legitimate means of war. The differentiation between 'good' and 'bad' chemical weapons was unfeasible and opened a loophole in efforts to stigmatize. The possibility that certain developments might open a space for legitimating technologies can be seen in more recent agreements. The 1993 Chemical Weapons Convention (CWC) prohibits the development, production, or retention of weapons that through their 'chemical action on life processes can cause death, temporary incapacitation, or permanent harm to humans or animals'. The CWC does, however, permit the use of agents for law-enforcement purposes, including domestic riot-control. Countries such as the United States have further interpreted this to mean that non-lethal riot-control agents could be permitted in certain non-conflict military situations (for instance, 'normal' peacekeeping) and that some uses of riot agents during armed conflict are not covered by the CWC (for instance, prison riots). There has been suggestion that ensuring that this interpretation is available has been a long-term aim of the United States.[43] The threat for Dando is that 'the immediate short-term advantages conferred by technological developments could endanger one prohibition regime and ultimately the complete set of arms-control regimes that the international community is trying to erect to restrain the proliferation of advanced weaponry'.[44] That the agencies of the CWC lack any independent intelligence functions to establish how weapons of law enforcement are being deployed and in what quantities further increases the likelihood of particular developments undermining calls for restraint.

The potential for this undermining effect also exists in relation to biological weapons. The Biological and Toxin Weapons Convention (BTWC) requires that

> Each State Party to this Convention undertakes never in any circumstances to develop, produce or stockpile or otherwise acquire or retain:
>
> 1. Microbial or other biological agents, or toxins whatever their origin or method of production, of types and in quantities that have no justification for prophylactic, protective or other peaceful purposes.

The broad and (one might think) fairly unequivocal wording of the BTWC means that those military biological agents discussed in Chapter 3 would not be permissible for warfare under international law. Much of the latest concern about the undermining of conventions through the strategic labeling practices relates to calmative agents and genetically engineered microbes.[45] Those pushing for arms controls often argue that, to function effectively, these agreements require trust-based coop-erative arrangements between parties; conditions that might be undermined by particular states being seen to manipulate their terms. By labeling calmative or malodorants as riot controls, or the situations of their use as 'military operations other than war', the prohibitions of the CWC and the BTWC might be side-stepped. US military legal advi-sors have suggested that calmatives would be generally acceptable under the BTWC, so long as they are classified as riot-control agents.[46] Likewise, fungi micro-organisms are being developed to destroy drug-producing crops. Such technologies are referred to as 'biological control' rather than 'biological warfare' means.[47] A greater danger for others is that the selective weakening of conventions and agreements, by making exemptions because of the non-lethality of certain weapons, might fundamentally undermine international law.[48] So, organizations such as the US Navy and the US Air Force's Armstrong Laboratory are genetically engineering highly selective microbial and biocatalysts that degrade material (such as runways or lubricants). Although such anti-material innovations have been ruled by the Naval Judge Advocate General as falling foul of the BTWC, they continue under the justifica-tion of being 'non-lethal'.[49]

The sorts of dispute that arise from the application of general agree-ments about warfare to particular situations are hardly unique to weapons controls. During the writing of the middle section of this book, media attention about the bombing of Afghanistan moved on to issues about the detention of those captured by US forces. International humanitarian law grants captured members of armed forces protection under the Third Geneva Convention. Just what protection those combatants captured in Afghanistan qualify for and how that ought to be determined have been the source of dispute between the United States and other countries and organizations. At stake in such debates is the possible undermining or reaffirming of international law due to particular classifications.

And as with the case of prisoners from Afghanistan, alternative views exist as to whether the upholding of international standards pertaining to non-lethal weapons is necessary or advisable. Here, attempts to set broad limits on weapons are taken to be both inappropriate and inef-fective.[50] As critics of technology in Chapter 6 drew on past experience to make determinations, so too do advocates – just with a different agenda in mind. Past experiences, such as medieval efforts to limit the crossbow or Cold War attempts to ban offensive biological weapons, are taken to indicate that controls are often unfeasible. Whereas for some the lack of confidence in the enforcement of controls is a potential cause of their failure and a reason to seek comprehensive measures, for others, past failures of controls are a reason for lacking confidence in such measures and thereby downgrading their relevance. General legal proscriptions are also taken as inappropriate because in certain situations a given weapon might prove highly effective or minimize casualties. As argued by Alexander and others, misconceived efforts to establish general controls on weapons stem from a failure to identify the source of the real problem (intent rather than technology) and a reliance on emotional arguments.[51]

Anti-Personnel Mines

Consider these issues in the debates about outlawing anti-personnel landmines. The 1997 Ottawa Treaty bans the use, development, stockpiling and transfer of anti-personnel landmines. The United States has not ratified the Treaty and is instead seeking to develop non-lethal mines capabilities such as rubber balls that incapacitate, calmative agents that tranquillize, nets and other devices that entangle, electroshock apparatuses that stun as well as a variety of more speculative incapacitation equipment based on sound and microwaves. Alexander supports such an approach when he argues:

> The fundamental legal problem that exists in use-of-force issues is that such problems are defined in terms of technology, when the real culprit is people and their intent. It is easy to develop an emotional argument against a technology or class of technology. There are an estimated 26,000 casualties from mines each year. Based on horrific pictures of men, women, and children suffering from traumatic amputations of their limbs, it is easy to generate support for a ban on the weapons that caused them …While organized with the best of intentions, [such anti-mine groups] focused on the wrong problem. The real problem was indiscriminate use of explosive devices that resulted in those maiming injuries and deaths. In many cases, it was mines that facilitated the tragedy. However, someone put them there.[52]

As such, the proper course of action is to evaluate the merits of particular weapons on a case-by-case basis, where the intent of those in question is the main arbiter of the likely appropriateness of particular deployments.

Following from discussions in previous chapters, it can be suggested that a characterization of some arguments as 'emotional' versus others as 'rational' is an attribution given to promote some positions over others. Rather than take such categorization at face value, it is prudent to consider the logic given. Those calling for a comprehensive ban of anti-personnel mines have justified this position on a number of grounds. As the humanitarian costs of such devices have far exceeded their military effectiveness, sweeping measures are necessary that stigmatize any such technology. Once allowances are made for particular exemptions, that stigma is diluted. Those in areas mined are unlikely to have their fear of movement reduced should they know of and believe in claims that the mines are probably not going to kill them.

Another reason for calling for a comprehensive ban and not making exceptions is that assessing likely deployments on their individual merits requires an unrealistic, rigorous policing of classification boundaries. Making some types of mine legitimate if used in prescribed ways in certain situations raises questions about how such standards can be enforced. For instance, in relation to existing measures, the Ottawa Treaty does not cover anti-vehicle mines. So, just where the boundary lies between anti-vehicle and anti-personnel mines is of significance. The group, Landmine Action, for instance, maintains that in some cases, a child running can exert a ground force in excess of the average anti-vehicle mine initiation pressure.[53] Such claims, then, set the stage for complex and detailed questions about what kinds of mine in which sorts of situation and with what frequency are liable to result in civilian casualties. Should 'non-lethal' mines be promoted as a substitute for existing personnel ones,[54] similar boundaries would have to be policed. Doubts about the feasibility of such a task, then, justify general prohibitions. Thus, the 'symbolic' aspects of arguments for certain bans are not simply emotional or subjective in character but relate to appraisals of the chances that individuals, organizations and technologies will function in the future in accordance with controls.

SIrUS – Science as Guiding the Way?

A recurring theme in the argument above is that while international law and arms-control treaties specify a number of general principles that prohibit certain types of weapons or their use, what such statements mean in practice is altogether more complicated. With the exception of blinding lasers, past formal international deliberations through the UN and elsewhere regarding whether a weapon was inhumane (such as cluster bombs or incendiary devices) only began after the death and injury toll mounted. For those concerned about the humanitarian implications of the methods of warfare, this situation has obvious severe drawbacks. In addition to the death and pain caused, by the time debates have started about a weapon's legitimacy, it has already become entrenched in the training and planning of at least some militaries. Under international law, vague criteria that blend legal and moral forms of argument – such as refraining from causing 'unnecessary suffering' – would then have to be evoked in a convincing fashion.

In an attempt to develop a forward-looking, 'clear and objective' scheme for assessment of conventional weapons – be they labeled lethal or non-lethal – members of the ICRC have offered medical criteria for defining 'superfluous injury' or 'unnecessary suffering' in the SIrUS project. Herein, superfluous injury and unnecessary suffering are 'determined by design-dependent, foreseeable effects of weapons when they are used against human beings and cause:

- specific disease, specific abnormal physiological state, specific abnormal psychological state, specific and permanent disability or disfigurement;
- field mortality of more than 25% or a hospital mortality of more than 5%;
- grade 3 wounds as measured by the Red Cross classification [skin wounds of 10 cm or more with a cavity];
- effects for which there are no well-recognized and proven treatments.[55]

Rather than pursuing a case-by-case form of governing weapons, these criteria are meant to provide a general classification scheme for all such devices: one based on their effects. The general criteria might mean banning a weapon because of its inherent effects or only banning its use in particular situations because of its foreseeable effects. It was not meant to prevent governments from seeking prohibitions against weapons on the basis of other considerations, such as public abhorrence or public interest criteria.

Criteria two and three were derived from hospital and casualty data on injuries sustained in conflict over the past 50 years, as well as a database of over 26,000 injuries collected by the ICRC during its missions in the 1990s to conflict areas such as Afghanistan, Cambodia and Sudan. In relation to criterion two, aggregate information gathered indicated that between 2.5 and 4.5 percent of casualties die after reaching medical facilities and between 18 and 25 percent of those wounded in the field die there. Severe grade 3 wounds have been inflicted in less than 10 percent of injuries from fragmentation or bullet injuries. These figures are taken as providing a foundation for devising a baseline against which individual weapons can be judged.

The ICRC has hosted a number of meetings of government, medical and legal experts to consider the SIrUS project criteria and encourage militaries to apply them to their weapon policies. Although the project was intended to provide an objective medical basis for assessment, its recommendations have not been accepted as such. As of yet, its uptake has been limited, though countries such as Australia have promised to factor the criteria into its policy decisions.

Opposition to SIrUS has stemmed from countries such as the United States, and this has centered on similar points raised throughout this book about the validity, conditionality and scope of claims. With regard to validity, the ICRC's data on conflicts has been said to be significantly unrepresentative of war conducted by advanced militaries, and thus the casualty rates are inappropriate even if taken as general guides.

Various assumptions underpinning the criteria can be questioned as well. The conclusions rely on a distinction between design and use-dependent effects. So, a weapon may transfer a certain amount of energy to the body because of its design, but its effects also vary depending on its use. The criteria specified are only meant to relate to design-dependent effects that are independent of use. Yet, concerns have been voiced about the feasibility and wisdom of separating out design and use-dependent effects. In relation to non-lethal weapons, for instance, as suggested throughout this book, many of their effects are said to be highly dependent on following strict guidelines. While disfigurement might not be a 'foreseeable' consequence of using a non-lethal weapon in accordance with operating rules (criterion 1 – for instance, as has been said of plastic bullets in Northern Ireland despite such effects), the likelihood and ability of users to follow rules is itself debated. In other words, the application of general prohibitions against weapons requires that assumptions be made about how they will be used. The general criteria also rest on additional contextual assumptions about the level and importance of medical treatment. Where medical facilities might be more or less advanced than the average facilities made use of in the ICRC data, death rates from particular weapons will differ and in turn affect what is deemed inhumane. Against this, members of the ICRC have argued that access to medical treatment in itself, rather than its quality, is the most important factor in determining the lethality of weapons.

As an example of a final line of criticism, the SIrUS criteria must be weighed against other humanitarian aims. Particular uses of force with a high potential for discrimination, such as sniper fire, would presumably be deemed inhumane because of their high fatality rate. Being general in character, the criteria are likely to rule out various employments of force that might otherwise have been deemed legitimate and preferable.

A practical problem is that if exceptions were made for particular uses of weapons in certain situations for any reason, this would undermine the strength of the criteria and bring a host of questions about whether in specific circumstances the standards should or should not apply. In any such deliberations, it seems likely that those wishing to oppose determinations of severe effects could vary the contextual situation under consideration, or the assumptions behind lethality claims (following the numerous examples given in this book), with some ease depending on the case at hand. While for practical purposes such debates might be resolved, it is likely that this would take time and effort, thereby diminishing the power of the criteria as providing objective and clear guidance. The difficulties to be overcome in applying SIrUS-like principles are not merely a matter of gathering sufficient empirical information or situating them in relation to wider legal or ethical issues. Rather, the basic problems are ones about determining what context is in question, what characteristics of weapons are relevant for consideration, and which boundaries are good enough to achieve particular goals. These cannot be resolved once and for all through analysis.

In sum, the SIrUS project can be subjected to various lines of questioning related to the reconciliation of the general criteria and particular situations. The aim of the points above is not to suggest, therefore, that any general standards for evaluations are necessarily unworkable. Rather, it is to acknowledge that, in any process of making determinations about what counts as a humane or inhumane weapon, the possibility for general criteria to provide an unambiguous basis for making determinations is limited. Plausible and convincing cases for the control of weapons can be made, but not from just describing the effects of technology. Those committed to finding deficiencies in criteria about the acceptability of weapons, rather than trying to make them work, will have little difficulty in doing so. The boundaries between what is necessary and unnecessary, as well as the appropriate context for evaluation, are chief areas for trying to undermine particular proscriptions.

MOVING FORWARD

Stepping back from the specifics pertaining to particular weapons, the previous argument has illustrated the contrasting lines of argumentation that exist for thinking about the relative merits of controls. Those alternative bases rest on assumptions and commitments about the institutions establishing and enforcing limitations. As has been suggested here, there are two key dynamics in debates at work in discussions about controls: the drawing of distinctions about and between weapons and the reconciliation of general and particular claims.

With regard to the first dynamic, the distinctions and classifications made of particular weapons are the focal point for debates about controls. Those offering different assessments of the merits of specific technologies attempt to marshal evidence to support particular classifications of the relative lethality or revolutionary potential of technologies. The position one takes on such matters affects evaluations of weapons' legitimacy. It is not just these determinations that are debated, but questions are raised about how categorizations can be policed in practice. In other words, at stake is how claims about what technologies are for can be confirmed or challenged.

Following the general 'analytical skepticism' approach outlined in Chapter 2, I want to refrain from marshalling particular claims about non-lethal weapons as a resource for making evaluations about their actual dangers in order to take the process of making distinctions as a topic of analysis. Key questions exist about how distinctions are 'used, when, by whom and to what effect'.[56] The next chapter in particular examines how distinctions are made in practice about CS sprays and what consequences follow. In taking this approach, I will further consider how statements about weapons often rest on implicit assumptions about the context features of their use. To the extent that general statements are given about the acceptability or unacceptability of non-lethals, their legitimacy is predicted on issues such as the contextual features of their use being irrelevant to the evaluation made. In this way, to talk about chemical sprays as being 'safe' or 'unsafe' can give way to more specific statements about what kinds of spray are safe and when.

With regard to the second dynamic, it has been argued that those commenting on the legitimacy of these weapons need to reconcile general and particular statements. For proponents, non-lethals are generally safe and those who employ them (in the West, at least) generally do so in an appropriate manner. In this way, non-lethals are meant to substantiate the relatively benign intent of those organizations deploying them and enhance their legitimacy. While there may be situations where unfortunate consequences arise or the weapons are abused, these are treated as aberrations. Against such presumptions, critical statements imply a rather different orientation. While non-lethals are said to be safe if used 'properly', such claims are based on questionable assumptions about how devices are used in practice. Furthermore, the effects of such technology are seen as varying considerably.

In simplified terms, critical positions struggle with the tension of making general and authoritative claims that can effectively justify limiting the deployment of non-lethals while trying to be responsive to the specificity and particularity of individual instances of use. Supportive positions highlight the importance and situated utility of specific instances of use but also wish to promote the generic merits of non-lethal force options for a wide range of circumstances.

There is a fundamental problem in any sort of evaluation: individuals are trying to find an appropriate reconciliation between making general claims that give some policy or other practical guidance and wanting to be responsive to the context-specific justifications for particular deployments.[57] Any attempts to establish a definitive assessment of non-lethals (whether it be positive or negative) is thus open to alternative criticisms: in the case of generalized assessments, it can be argued that crucial but contingent variables have been suppressed. So, on normal and healthy people a weapon might be relatively safe, but what about those with medical conditions? International laws might deem that types of weapon are 'justly condemned', but that in certain cases they might prove valuable and more appropriate than other options. In the case of situation-specific assessments (in this instance, has it been used appropriately or not?) the counter can be made that nothing of much general applicability is being said. If such weapons are seen to work well in a specific case, what does this imply for their wider uptake? Attempts to establish definitive assessments of non-lethals are thus open to alternative criticisms that crucial but contingent variables have been suppressed (as in the case of general claims), or that nothing much in the way of general relevance is being offered (as in the case of specific claims).

The disagreements and lack of determinacy associated with the effects of non-lethals makes this reconciliation between the particular and the general especially tension-ridden. While this is a problem for those wishing to make authoritative statements about the rights and wrongs of non-lethals because it calls into doubt the legitimacy of their claims, I want to suggest here that it can also be a source of insight. One way of usefully working with the indeterminacies without trying to settle them is to consider how they are distributed and the implications of this.

Consider these points in relation to a specific issue. Throughout this book, it has been argued that determinations of what a weapon can and cannot do and what it is for are tied up with questions about responsibility. In considering questions about legitimacy, responsibility and distributions of indeterminacy, I want to draw on work outside of the security studies. Nick Lee[58] has examined how the distribution of ambiguity surrounding the ability of children to speak for themselves relates to questions of institutional legitimacy. Within adult institutions, children are ambiguous creatures. It is rarely clear who is supposed to speak for a given child, whether that is the child herself or himself, parents or authority figures. In practice, the moment in which these ambiguities

167

should be resolved is frequently deferred. For instance, the status of children as competent actors in legal proceedings is often disputed. There are understandable and persistent difficulties in making generalized policies about the status of children as competent speakers (for instance, by establishing a certain minimum age). Yet, for institutional actors' decisions to be more than merely ad hoc, some degree of generality must be entailed. The ambiguous status of children means that legal institutions often defer taking the time to decide on whether a particular child is competent to give evidence until actually in the courtroom; a process that often brings significant stress for the child. In such conditions, Lee argues that a key concern 'is not to answer the question of children's status, but to examine how that question accompanies children in their passage through various social orders'.[59] How ambiguities related to the status of children are distributed, where they must be resolved and by whom are key questions.

A similar approach can be taken in examining disputes about the status of weapons. Evaluations of non-lethals are at once adjudications about the status of technology made in relation to concerns about the legitimacy of those pronouncing them. Following Lee, the focus can become one of asking how the treatment of 'ambiguity' – what I consider here to be the indeterminacies, disputes and conditionalities – accompanies non-lethals and secures or undermines their legitimacy. A consideration of these sorts of institutional concerns then provides a basis for asking 'which kinds of versions win the day, in what circumstances, and why'.[60] It also enables us to see how responsibility is allocated in contested situations. Chapters 8 and 9, in particular, take up these themes.

SUMMING UP

In disputes about how and whether non-lethal weapons ought to be controlled or limited, matters of empirical evidence, trust in organizations, and past experience mix readily. In suggesting particular regulation measures, questions exist about when it is justified or not to close down deliberations and take as fixed the characteristics of weapons or the likely actions of their users. To the extent that export controls are applied to non-lethals, determinations of the appropriateness of transfers rest on judgments about such things as the likely employment of weapons by recipients, the merits of the weapon, or both. Claims about the legitimacy of past or future uses of non-lethal weapons are laced with determinations of agency and notions about the credibility of different types of argument. Where certain prohibitions are suggested or agreed, the application of general principles to specific cases provides a source for continuing negotiation. Governing procedures are fragile and potentially open to challenge from a variety of directions. 'Get-out' clauses are built in or invented to provide for the flexible adaptation of policies.

Beyond painting this picture, this chapter has suggested key areas for the analysis of non-lethal weapons. Typically, the terms of the debate about the merits and effects of non-lethals center around circumscribed considerations, for instance whether they are 'actually' 'non-lethal' or whether they may produce 'lethal' consequences. It would be possible to try to clarify the terms of the debate in the hope of shoring up a particular reading. However, two suggestions for alternative lines of investigation were presented in this chapter: taking distinctions made as a topic of analysis and paying attention to the distribution of indeterminacies in formulation of notions of responsibility. It is to these issues that the next chapter turns.

NOTES

1. Amnesty International, 'Assisting Wrongful Acts', *The Terror Trade Times*, June 2001, 1.
2. See for instance Hagelin, B., Wezeman, P., Siemon, T., Wezeman, W. and Chipperfield, N., 'Transfers of Major Conventional Weapons', *SIPRI Yearbook 2001: Armaments, Disarmament and International Security* (Oxford: Oxford University Press, 2001); Amnesty International (Irish Section), *MSP Companies and Transfers* (Dublin: Amnesty International Ireland, 2001); Saferworld, *The EU Code of Conduct on the Arms Trade* (London: Saferworld, 2001); Human Rights Watch, *Money Talks* (Washington, DC: Human Rights Watch, 1999); Gabelnick, T., Hartung, W. and Washburn, J., *Arming Repression* (Washington, DC: World Policy Institute and the Federation of American Scientists, 1999); and Lumpe, L. (ed.), *Running Guns* (London: Zed, 2000).
3. See e.g. British Defence, Foreign Affairs, International Development and Trade And Industry Committees, *First, Second, Third, Fourth Reports*, 2 February (London: HMSO, 2000).
4. Hartung, W. and Moix, B., *Deadly Legacy* (Washington, DC: World Policy Institute, 2000) and Mitchell, E., *UK Arms Exports to Zimbabwe* (London: Campaign Against the Arms Trade, 2000).
5. Wood, B. and Peleman, J., *The Arms Fixers* (Oslo: International Peace Research Institute, 1999).
6. Mariani, B. and Urquhart, A., *Transparency and Accountability in European Arms Export Controls* (London: Saferworld, 2000).
7. Amnesty International, 'Zambia', *The Terror Trade Times*, October 1999, 1.
8. Amnesty International, 'Tear Tracks', *The Terror Trade Times*, October 1999.
9. Amnesty International, *Israel/Occupied Territories and the Palestinian Authority* (London: Amnesty International Secretariat, 1998).
10. Krau, N., 'France and Germany Stop Arms Sales to Israel', *Ha-aretz*, 17 December 2000.
11. BBC News, 'France Refuses Israel Tear Gas Grenades', *BBC News*, 19 December 2000.
12. In any case, Israeli security forces do not seem to have been left wanting for tear gas because of American supplies. See Risk, R., 'Death in Bethlehem, Made in America', *The Independent*, 15 April 2001.
13. Opall-Rome, B., 'Israeli Gunship Crews Train for Assassination Missions', *Defense News* 26 November–2 December 2001, 30.

14. Lewer, N. and Feakin, T., 'Perspectives and Implications for the Proliferation of Non-Lethal Weapons', *Medicine, Conflict, and Survival*, 17, 3 (2001), 272–85.
15. South Africa has also gained the ability to produce CR. See Centre for Conflict Resolution, 'Inside Track', *Track Two*, 10, 3 (2001).
16. Abel, P., 'Manufacturing Trends', in L. Lumpe, *Running Guns* (London: Zed, 2001).
17. Amnesty International, 'Breaking the Power of the Brokers', *The Terror Trade Times*, October 1999.
18. Amnesty International, *Stopping the Torture Trade* (London: Amnesty International, International Secretariat, 2001).
19. Amnesty International, *Arming the Torturers* (London: Amnesty International, International Secretariat, 1997), 1.
20. This amount only refers to non-military exports; see Bureau of Export Administration, *The BXA Foreign Policy Report to Congress*, 18 January (Washington, DC: BXA, 2002). The policy of denying the public release of information was provided by Mortimer, K., personal communication with author, 25 February (Washington, DC: BXA, 2002).
21. Amnesty International, *Cruelty in Control?* AMR 51/54/99 (London: Amnesty International, 1999).
22. Committee Against Torture, *Conclusions and Recommendations of the Committee against Torture*, 15/05/2000 A/55/44 (Geneva: CAT, 2000).
23. See Alexander, J., *Future War* (New York: St Martin's Press, 1999), 185–6.
24. Much debate has taken place in Europe. In April 2001, for instance, the European Union's General Affairs Council agreed 'Guidelines to EU Policy towards Third Countries on Torture and Other Cruel, Inhuman or Degrading Treatment or Punishment'. The Guidelines call for European countries to encourage others to prevent the use, production and trade of equipment which is designed to inflict torture or other cruel, inhuman or degrading treatment or punishment and prevent the abuse of any other equipment to these ends.
25. See www.amnesty-usa.org/news/2001/usa08012001_2.html
26. Amnesty International (UK Section), *Submission to the NIO Steering Group on Recommendations 69 and 70 of the Patten Report*, February (London: AIUK, 2002).
27. Rejali, D., 'Technological Invention and Diffusion of Torture Equipment', presented at the International Sociological Association Conference Montreal, August 1998, see http://www.reed.edu/~rejali/articles/electric.html, and Rejali, D., *Electric Torture*, February 1999, see http://internationalstudies.uchicago.edu/torture/abstracts/dariusrejali.html
28. In an effort to present a proper framing for examining the legitimacy of electroshock, Rejali presents the origin of such weapons as more complicated than either story above suggests, where their invention took place through a mixture of international military, industrial and psychiatric practices. The rise of these weapons as instruments of force and torture is for him bound up with processes of international and national scrutiny. The difficulty of tracing perpetrators or substantiating claims of abuse from electrical devices means that their uptake informs us more about democratization than about authoritarianism.
29. Bleetman, A. and Steyn, R., *The Advanced Taser*, 17 December 2000, 18–19.
30. For an introduction to some of the issues see O'Brien, W., *The Conduct of Just and Limited War* (New York: Praeger, 1983) and Walzer, M., *Just and Unjust Wars* (New York: Basic, 1997).
31. See for instance Coppernoll, M.A. and Maruyama, X., 'Legal and Ethical Guiding Principles and Constraints', in *Proceedings of NDIA Non-Lethal Defense III Conference*, 25–26 February 1998 (National Defense Industrial Association Conference, 1998).
32. Lewer, N. and Schofield, S., *Non-Lethal Weapons* (London: Zed, 1997), 86.
33. See Fidler, D., '"Non-Lethal" Weapons and International Law', *Medicine, Conflict and Law*, 17, 3 (2001), 194–226, and Lewer and Schofield, *Non-Lethal Weapons*, Chapter 4.
34. Fidler, '"Non-Lethal" Weapons and International Law', 197.
35. Doswald-Beck, L., 'New Protocol on Blinding Laser Weapons',*International Review of the Red Cross*, 36 (1996), 272–99.
36. Warford, J., 'The Chinese Type 98 Main Battle Tank', *China Defense.com* – see

www.china-defense.com/armor/Type98/type98_3.html

37. Kennedy, T., 'Israel Acquiring Banned American Laser That "Melts Eyeballs" of Enemies', *The Washington Report*, July 1996, see www.washington-report.org/backissues/0796/9607036.htm

38. Human Rights Watch, 'HRW Questions U.S. Laser Programs As Blinding Laser Weapon Ban Becomes International Law', *Human Rights Watch*, Press Release, 29 July 1995.

39. Dando, M., *A New Form of Warfare* (London: Brassey's, 1996), Chapter 6.

40. See Goldblat, J., 'Are Tear Gas and Herbicides Permitted Weapons?' *New Scientist*, April (1970), 13–16.

41. Parks, H., 'Classification of Chemical and Biological Warfare', *University of Toledo Law Review*, 13 (1982), 1145–72.

42. Ackroyd, C. et al., *Technologies of Political Control* (London: Pluto, 1980), 216.

43. See Dando, M., 'Future Incapacitating Chemical Agents', in N. Lewer (ed.), *The Future of Non-Lethal Weapons* (London: Frank Cass, 2002).

44. Dando, *A New Form of Warfare*, 188.

45. See The Sunshine Project, *Non-Lethal Weapons Research in the US*, July (Washington, DC: The Sunshine Project, 2001).

46. See Coppernoll and Maruyama, 'Legal and Ethical Guiding Principles and Constraints'.

47. See Sunshine Project, *An Introduction to Biological Weapons, their Prohibition, and the Relationship to Biosafety* (Washington, DC: The Sunshine Project, 2002).

48. For a discussion of this see Fidler, ' "Non-Lethal" Weapons and International Law'.

49. For a discussion of this see The Sunshine Project, *US Armed Forces Push for Offensive Biological Weapons Development*, 8 May (Hamburg: The Sunshine Project, 2002).

50. See Alexander, *Future War*, Chapter 19, and Council on Foreign Relations *Nonlethal Technologies* (Washington, DC: Council on Foreign Relations, 1999).

51. See as well Smith, R., 'Reducing Violence', Proceeding from *Security Systems and Nonlethal Technologies for Law Enforcement*, 19–21 November 1996, Boston, MA (Bellingham, WA: SPIE, 1997), 27.

52. Alexander, *Future War*, 197.

53. Landmine Action, *Civilian Footsteps* (London: Landmine Action, 2001).

54. See German Initiative to Ban Landmines and Landmine Action, *Alternative Anti-Personnel Mines* (London: Landmine Action, 2001).

55. International Committee of the Red Cross, *The SIrUS Project* (Geneva: International Committee of the Red Cross Publications, 1997), 8.

56. Grint, K. and Woolgar, S., *The Machine at Work* (Cambridge: Polity, 1997), 68.

57. For some exemplary analyses about the general topics see Davis, K., *Discretionary Justice* (Baton Rouge, LA: Louisiana State University Press, 1969) and Garfinkel, H., *Studies in Ethnomethodology* (Cambridge: Polity, 1994).

58. Lee, N., 'The Challenge of Childhood', *Childhood*, 6, 1 (1999), 455–74.

59. Ibid., 464–5.

60. Grint and Woolgar, *The Machine at Work,* 368.

PART III:
CASE STUDIES

8

CS Sprays in Britain

The chapters of Part II surveyed various aspects of non-lethal weapons in the use of force: their effects and employment, and the prospects for their control. It was stressed throughout that notions about what technologies are for and what they can do were bound in an understanding of their 'context'. Thus, the separation between examining areas of effects, use and control undertaken in the last three chapters has been artificial and in some respects unhelpful. The chapters in Part III provide a detailed analysis of particular deployments of non-lethals so as to more fully illustrate the issues and themes raised in previous chapters. As will be argued, it is only through such detailed analysis that one can fully grasp the potential and problems associated with this set of technologies.

This chapter examines the uptake of CS chemical sprays in England and Wales.[1] In some respects, the deployment of CS sprays provides somewhat of a 'best case' scenario for achieving the potential of non-lethality. As opposed to many of the types of highly charged public-order events discussed previously, everyday policing does not evoke the same degree of controversy. In addition, the supposed 'consensual' nature of policing in Britain (at least in comparison to other parts of the world) and the previous experiences with similar 'off-the-shelf' sprays in other countries should have meant that their introduction was fairly unproblematic. Also, the unarmed status of many police officers gave an additional justification for the spray's introduction in Britain. And yet, for many of the same reasons, the spray also proved controversial. So various commentators expressed concern that their deployment represented the imposition of 'alien' and aggressive styles of policing.[2] In terms of the analysis of this book, this case is an important one for another reason: the introduction of the CS sprays in Britain was the first deployment of a non-lethal weapon that I studied. The analysis of this case has played an important role in informing the overall framing and conclusions given here.

A story of the uptake of CS sprays could focus on various facets (their history, training procedures, use in specific field situations). This chapter considers those aspects of their deployment related to the themes of the legitimacy of the use of force. A number of questions are addressed along these lines: How have claims about the relative safety of the sprays been substantiated and employed? How are notions about their use and safety mutually defined? How are notions about blame and responsibility negotiated? Following the conceptual discussion at the end of the last chapter, distinctions made about weapons and the distributions of indeterminacies in formulations of responsibility are used to develop an understanding of the constitutive relation between rules and practices within organizations.

THE INTRODUCTION OF CS SPRAYS

While various policing bodies in Britain were pressing for the adoption of some form of chemical incapacitant since the early 1990s, just what chemical-based spray should be brought in and what governing procedures established were sources of disagreement. Despite the wide-scale adoption of pepper sprays in the United States, as mentioned in Chapter 5, fairly persistent concerns have been voiced regarding the lack of knowledge about their health effects, the possibility for overexposure, and the variability of effects due to pre-existing medical conditions. Based on such concerns about safety, in 1994 the UK Police Scientific Development Branch (PSDB) – the primary safety-testing organization for the police – concluded that pepper spray posed too great a risk. The Association of Chief Police Officers (ACPO), acting in an advisory fashion to the police in liaison with the PSDB, instead approved CS sprays with the solvent methyl isobutyl ketone, similar to those used by the French Gendarmerie at the time.

CS sprays were trialed and later approved for use in English and Welsh forces in 1996. The device was said to provide an effective option in handling unruly individuals from a distance, to deter assaults against officers, and to reduce excessive-force complaints against the police.[3] The sprays were generally portrayed as an intermediate option between the baton and firearms, which could prevent the need to routinely arm British officers.[4]

ACCEPTABLE SAFETY?

The safety of the sprays has been a matter of discussion throughout the trial and during their deployment. Those wishing to substantiate safety claims have pointed toward various testing procedures and field experiences. Several major evaluations of the sprays have been conducted and these taken as vindication of safety. Despite government agencies' assurances of the benign character of the incapacitants, various counter-arguments have been offered. Many critical comments have drawn on the testing procedures and field experience for evidence, though in order to reach opposed conclusions. If the debate about their merits is posed in terms of 'Are CS Sprays Safe?',[5] then the answer is clearly no. Arguably, chemicals designed to cause pain and irritation will never be free from risk. The more difficult questions and the ones occupying most commentators include: How safe are the sprays? Compared to what other options should their safety and effectiveness be judged? What evidence has been deemed adequate to comment on these issues? Has everything been done that is reasonably possible to ensure that the effects are minimal?

Addressing these questions requires delving into the nitty-gritty of statements. This chapter undertakes such an endeavor, with a view to addressing two (seemingly) simple issues: what have been the deleterious effects of the sprays and why have these taken place? As I argue below, this requires tracing the movement of statements between organizations to see how facts are constituted. In these circumstances, there are pressing questions about how one cuts through disputed socio-technical assemblages in the hopes of offering particular evaluations of what has happened and why.

Like a Drug?

As outlined in Chapter 5, the Himsworth Committee report into the deployment of CS gas in Northern Ireland in the late 1960s and early 1970s has provided much of the regulatory framework for the approval of riot-control agents in Britain. Suggesting various uncertainties about the effects of CS, the Committee offered two central procedural recommendations: riot-control agents should be regulated as being more akin to drugs than to weapons for their approval, and the evidential basis of safety claims should be published promptly in medical and scientific literature.

The acute effects of CS can include a severe burning discomfort, pain in breathing, the production of tears, coughing, vomiting, erythema, dyspnoea and involuntary eye winking. It is associated with conditions such as the constriction of airways in asthmatics. Since the time of the Himsworth Committee, the possibilities of various chronic effects have been raised. These include permanent lung damage at comparatively low doses, shortness of breath, second-degree burns, chemical sensitiza-tion, chromosomal damage, skin blistering, severe dermatitis and the triggering of heart failure.[6]

Throughout the uptake of CS sprays, certain government and police officials have stated that the recommendations of Himsworth have been followed and that the sprays are relatively benign.[7] Commenting on the appropriateness of the sprays, Lord Williams argued that

> CS has been scientifically evaluated to a level similar to that which would be required for a new pharmaceutical drug and has been found not to present any significant threat to human health.[8]

Seemingly reiterating this, former Home Office Minister Alun Michael said that

> CS spray has been scientifically tested to a level similar to that which would be required for a new pharmaceutical drug, and there is no evidence that it poses any significant threat to human health.[9]

The specifics of the approval procedures, however, have not been made widely available.[10] As opposed to licenced pharmaceutical drugs in Britain, the sprays did not go through the regulatory testing procedures of the Medicines Control Agency. In this situation full of unknowns, early criticisms of the sprays in 1996 drew attention to the lack of public knowledge about the procedures and also leaked police guide-lines that cautioned about the possibility for eye damage.[11] After some evidence of injuries, in 1998 the medical journal *The Lancet* called for a moratorium on the deployment of the sprays until details were published regarding the basis for evidence about their safety.[12] The Sussex, Northamptonshire and Nottinghamshire police forces refused to issue the sprays, citing doubts about the rigorousness of the proce-dures as known to them.[13] In 2001, without the backing of the Home Office, Sussex began trials of an OC-related spray called PAVA. The failure of the Home Office to give approval here stemmed from PAVA's not being tested in a manner similar to that required for a pharmaceu-tical drug. Since the testing procedures were not stipulated for CS sprays, just what this phrase meant in practice has been a topic of some uncertainty. The debate has centered on the testing required for what counts as 'akin to a drug' – whether that be by the procedures required for drug regulation today or by those in place in the early 1970s when the Himsworth Committee came to its conclusions.

Eventually, details of the approval process for CS sprays have become known to some. In 1999, Matthew Taylor (a member of the UK House of Commons) wrote to the Home Office and the Department of Health asking for clarification of the basis of evidence by citing the Himsworth Committee recommendations. Those details were passed on to me through an investigative journalist. Figure 8.1 presents extended extracts from these letters.

As set out, the justification for safety claims rested on a series of reports by the Chemical and Biological Defence Establishment at Porton

---

FIGURE 8.1
EXCERPTS FROM CORRESPONDENCE WITH MATTHEW TAYLOR MP
ELABORATING THE CS SPRAYS APPROVAL PROCESS

Paul Boateng MP, Home Office Minister, Letter dated 8 January 1999[14]

Your letter also refers to the [CS spray] solvent Methyl Iso-butyl Ketone, known as MIBK. MIBK has been subjected to extensive testing over many years. Information from these studies is currently being reviewed by the 'Independent Standing Committees on Chemicals in Food, Consumer Products and the Environment' which are supported by the Department of Health. These committees regularly advise the Department of Health on a number of issues and were not set up specially to review CS or MIBK.

Expert opinion provided to the Home Office Police Scientific Development Branch has stated that the toxicity of CS and MIBK will not be any greater than the sum of the individual toxicities of CS and MIBK. In this respect the combination of CS and MIBK is unlikely to pose a long-term hazard to health.

Finally, you ask for this Department to release reports of research into the use of CS. It is not entirely clear to which reports your letter refers. However, I can inform you that in 1994 the Chemical & Biological Defence Establishment (CBDE) was commissioned to produce a report on this subject. The report is unclassified and copies have been placed in the Library of this House. Furthermore, another report conducted by CBDE on CS solvents in July 1996 and a review of other potential solvents in November 1997 was also placed in the House of Commons Library.

Tessa Jowell MP, Department of Health Minister, Letter dated 30 March 1999[15]

As regards the choice of solvent for the spray, in March 1995 a view of Department of Health officials that methyl isobutyl ketone (MIBK) was a preferable solvent was conveyed to the Home Office Police Scientific Development Branch (PSDB). This view was made in the context of some public concern about the carcinogenicity of methylene chloride [MC] in experimental animals and also included the solvents MIBK, fluorocarbons and polyethyleneglycol 300. This was not a formal review of the alternatives, but was based on consideration of the data available in standard toxicology reference books. No specific new research was carried out by the Department, but subsequently HM [Her Majesty's] Government, through the Home Office PSDB, commissioned work by DERA Porton Down to review the scientific literature on the toxicology of methylene chloride and MIBK, and interdepartmental consultation on the issues was extensive. A copy of the review by DERA Porton Down has been deposited in the Library of the House.

Down, along with other input from government agencies. The solvent for the sprays, MIBK, was chosen as a safer comparison in relation to existing alternatives through scanning textbooks and, later, literature reviews.

'Protecting by Scientific Understanding'[16]

This section analyzes the aforementioned reports as well as other safety claims about CS sprays in relation to critical concerns voiced. Despite the concerns raised above about CS, the 1994 Porton Down report concluded that CS was the safest agent for liquid aerosols. CN and CR had major known hazards and, citing the Salem et al. report mentioned in the fourth part of Chapter 5 (see p. 104), OC was said to have lingering safety doubts. The authors (Rice and Jugg) stated that even repeated exposures to small dosages of CS over a prolonged period of time would not cause ill effects.

In contrast to the terms of debate set out in the 1994 report, many of the critical points made relate not so much to their deployment as an *aerosol* but to that as a *spray*. The latter are said to entail far higher levels and alternative types of skin and eye problems than the former. While aerosols release mainly gas particles, sprays discharge gas and liquid streams that give much-enhanced exposures. Therefore, Gray and others have argued that the findings of the Himsworth Committee about the relatively benign status of CS (as a gas or smoke) should not apply.[17] Thus, considering exactly what is being praised or criticized in statements is a matter of some importance. While the quotes given above from Lord Williams and Minister Michael are quite similar, they differ in referring to CS in general and CS as a spray. In the vast majority of supportive medical and government evaluations, CS as a gas rather than as a spray has been commented on.[18]

Likewise, just which 'CS spray' is under evaluation is another basis for disagreement. While CS sprays have been used for some time, the specific version in Britain was said to be far too high in its concentration. As Jones argued, the 5 percent (weight/volume) CS solution releases 5 centiliters of fluid per burst, where elsewhere a 1 percent solution releasing a 1 centiliter burst is common. In other words, someone exposed to the former would receive 25 times as much irritant.[19] While it has been reported that the French Gendarmerie did not systematically monitor the effects of the 5 percent spray,[20] cases of severe dermatitis and extensive blistering lasting for several days have been attributed to CS sprays in France.[21]

The specifics of the formulation are particularly important in the case of CS sprays because of concerns about the solvent MIBK.

Although much of the initial critical attention regarding the sprays centered on CS, later this shifted to MIBK and MIBK in combination with CS. With its focus on aerosols, the 1994 Porton Down report did not consider solvent options in any detail, other than to note that while methylene chloride (MC) was (and still is) the solvent for CS canisters for the British military, it was unacceptable for civilian situations.

From the parliamentary correspondence cited in Figure 8.1 regarding the approval process, it might be assumed that the July 1996 and November 1997 Porton Down studies reaffirmed the general safety of the solvent. However, this was not the case. The 1996 report, conducted by Rice, Dyson and Upshall (completed one month before the Home Office approved the CS sprays), compared MC and MIBK as possible solvents. Although having earlier ruled out MC as unacceptable for civilian use, the authors argued that it posed 'significantly reduced risk' compared to MIBK and 'strongly recommended' replacing the latter with the former.[22] This conclusion was overruled by the textbook review conducted by the Department of Health mentioned in the Jowell letter. From limited operational experience, in the 1996 report MIBK was attributed with causing 'delayed symptoms [such as reddening, peeling and blistering of the skin] commencing approximately six hours after exposure and lasting for several days'. In this respect, the Porton Down conclusions differ markedly from the original French manufacturer's claims that '[i]n effect, the solution used has been retained for its harmlessness to the skin, mucous membranes and especially the eyes'.[23]

The 1997 report (conducted by Rice, Jones and Stanton) reviewed available solvent alternatives and concluded that '[s]everal of these solvents are either confirmed or suspected carcinogens with associated mutagenic potential and clearly do not represent safe alternative solvents; we can, therefore, exclude methyl isobutyl ketone [MIBK] …'.[24] Although government ministers cited the 1996 and 1997 studies in 1999 as key pillars of the testing process and therefore tacitly as support for claims about safety, their recommendations suggested an evaluation at odds with the official one. In the only public comment on the Porton Down studies I was able to find (or at least repeat[25]), in July 1999 the Home Office said that the 1997 study was only a theoretical exercise that did not reflect actual operational exposures.

According to reports, neither members of ACPO nor the Police Federation (the rank-and-file police 'union') had seen the Porton Down studies until late 1998.[26] After media reports about these studies in late 1998, the safety of the sprays was referred to the independent expert government advisory Committee on Toxicity (COT) in Food, Consumer Products and the Environment, along with the assistance of the Committees on Carcinogenicity and Mutagenicity. The study of these committees consisted of a further medical literature review of toxicity, mutagenicity and carcinogenicity, in addition to a survey of limited operational data and flow rate studies of CS sprays.

As expressed by the weekly *Police Review* magazine editor, COT members faced a basic tension in the evaluation: endorse the safety of the sprays against the advice of Porton Down and other possible concerns about CS sprays and the committee's findings might later be called into question; condemn the sprays and the litigation possibilities from the public and the police would be enormous.[27] In the end, though, COT chose neither to condemn nor to vindicate. Its findings both acknowledged cause for concern and reassured those concerned.

In 1999, COT found that 'the *available* data did not, in general, raise concerns regarding the health effects of CS spray itself ... It must be noted that no comprehensive investigation of the effects of CS spray in humans was available, nor has there been any systematic follow-up of individuals who have been sprayed with CS spray' (emphasis in original).[28] Against concerns expressed about the possibility of severe dangers posed by the combination of CS and MIBK – for instance the carcinogenic risks of repeated exposures to the skin and airways – the Committee 'noted the sparsity of data on the combination of CS dissolved in MIBK. There are no data available on the metabolism, kinetics, acute toxicity, or skin irritancy of CS when administered in MIBK solution.' Stated differently, there was not enough information to allow for an evaluation of the safety of the particular sprays being used. 'Concern' was expressed for certain susceptible groups, such those with asthma, chronic obstructive airways, hypertension, other forms of cardiovascular disease, or those taking neuroleptic drugs. No effort was made to quantify or otherwise weigh the likely risks or benefits of the incapacitants. Despite the literature review format of the COT review, no mention was made of the findings of the Home Office-commissioned Porton Down reports. This was justified to the author on the basis of these reports being literature reviews in themselves and that the COT only drew on primary source material.[29]

As with the other reports about chemical sprays mentioned in Chapter 5, the COT review did not try to assess whether the sprays were being used in the prescribed manner deemed necessary (see below), though it acknowledged that its conclusions were predicated on the adherence to rules. The probability of developing dermatitis, for instance, would be greater from multiple-shot exposures. Stating that there were numerous uncertainties about effects, they suggested that those sprayed should be monitored for delayed injuries.

As with so many of the cases of non-lethal weapons surveyed in previous chapters, and with a situation which is arguably endemic to regulatory sciences more generally, medical and scientific evaluations were not able to resolve the 'facts of the matter' or even specify the risks entailed.[30] Not surprisingly, the indeterminate and qualified character of the COT report gave grounds for alternative interpretations. The Newham Monitoring Project suggested that the findings were 'entirely irrelevant because they do nothing to address the safety of the actual CS spray canisters issued to officers' and that in any case 'insufficient research' existed to prove safety claims.[31] A Home Office press release welcomed the COT report findings, stating it gave 'no reason to prevent the police service from using CS spray'.[32] While not inaccurate, the statement also does not allude to the conclusions about the 'sparsity' of knowledge about CS and MIBK or the qualified status of the COT report's findings.

'Building a Safe, Just and Tolerant Society'[33]

Although conclusive evaluations of safety might not have been forth-coming from medical and scientific evaluations, many of the hazards associated with CS sprays should have manifested themselves by now in their practical deployment. Besides predictive testing, the regulation procedures for drugs also include so-called post-licencing surveillance of adverse reactions. Although waiting for and cataloguing injuries is not the most desirable form of determining safety, in the case of CS sprays, experiential information on the number, extent and cause of injuries should help determine safety. And yet, as will be argued, data on injuries is either absent, inconclusive, or publicly inaccessible.

Countervailing claims have been made regarding their dangers since the start of employing the sprays. Internal police trials in 1995 were stopped after injuries were sustained by two Surrey officers, members of the PSDB, and a Metropolitan Police inspector.[34] The last suffered 40 percent and 50 percent corneal eye burns. Officers Gregory and Knill quote from a meeting of PSDB, Porton Down, and Department of Health officials that located the reason for this to:

1. Trainer Error – The spray was delivered too close to the face. This is a problem that will easily be replicated in stressful operational usage. It has also been stressed by a number of eminent sources that the 'jet' system specified by the PSDB can result in 'Hydraulic Needle' damage if delivered at a distance of less than 1 metre (the pressure of the jet injects the eye with the CS crystals and MIBK).

2. Aftercare was not as recommended.[35]

In other words, the fault lay with the individuals 'behind' the sprays. The police trainer in question disagreed, and became an adamant critic of them. In March 1996, the Home Office began the official operational trials of the spray in 16 forces throughout England and Wales. As part of this, an official appraisal of safety and effectiveness was commissioned. With regard to officer and public injuries, the review compared injuries from trial and control locations. Little difference was said to exist in the rate or character of officer injury across these areas, though the cause of injury was said to differ. Cross-contamination to officers from use of the sprays (due to, say, a change in wind direction or touching affected surfaces) were incurred in 78 percent of cases of the use of the sprays. It is not clear if any attempt was made to record delayed complications of those sprayed, though members of the Home Office had some knowledge of such effects. The review was also unable to comment on the effects of multiple exposures. Trial figures suggested that officers were likely to use the sprays once every 32 months and experience some form of cross-contamination once every six months. This rate of exposure means the negative effects associated with multiple exposures would only become apparent over many years in many cases.

Since the introduction of the sprays, a number of medical studies have been undertaken based on limited surveys of affected officers and members of the public. Consistent with the points made above, cases of eye damage,[36] delayed drying, flaking and blistering of the skin,[37] and allergic sensitization[38] have been said to take place. The possibility of extrapolating from these cases to determine rates of injury are limited, as the studies documented anecdotal cases that become known to particular individuals. Such studies by medical practitioners complement accounts in the media and newspapers about serious injuries[39] and the possible contributory role of CS sprays in deaths,[40] the latter topic itself deserving detailed analysis.

Comprehensive clinical data is not available on injury rates. The National Poisons Information Service undertakes a general surveillance function of the use of riot-control-agent injuries. These centers advise and collate information on poisons-related exposures where medical practitioners are uncertain as to the proper course of action. In 1997, the London branch recorded 597 patient reports of injuries with riot-control agents, though this included injuries from all forms of incapacitant sprays (OC, CS, CN), both those used by the police and those employed (illegally) by members of the public.[41] Delayed effects were catalogued (those taking place after six hours of exposure), though these were hardly surprising given the 1996 Porton Down study mentioned above. The injuries included cases of severe skin reactions attributed to police CS sprays.[42] Because the Poisons Service is only responsive to cases where practitioners are not able to administer care, the data is incomplete and of only highly speculative generalizability.

Arguably, officers are the most at-risk sub-population for sustaining ill effects due to the likelihood of cross-contamination and the possibility of experiencing repeated exposures. Yet, in relation to this group, those searching for definitive information about the extent and nature of injury from CS sprays are confounded by the lack of data and doubts about the validity, conditionality and the scope of claims, as well as by questions about the credibility and bases for identifying risks.

The Home Office and other departments do not hold figures on the number of legal proceedings, or compensation awarded to officers or members of the public, from CS sprays.[43] In the case of officers, compensation awards might come from civil cases or workplace compensation measures. Personal attempts to obtain information on the number of compensation claims to officers from the Police Federation were met with no precision of figures but elicited a response that the number of incidents was 'insignificant considering [the spray's] overall use'.[44]

Even if overall compensation claims to the officers were made available, there is reason to doubt just how representative these figures could be due to legal advise given to officers post-COT. This advice given to officers and local Police Federation branches states that there are no reasonable prospects for suing the police on the basis of the sprays being 'unsafe'. Officers are still able to get support from the Federation in the instances of injuries from negligence (say with regard to training) but not because of 'inherent' safety defects of the sprays.

According to the solicitor handling the civil cases of compensation for police officers, (however unfortunately) the COT report had become the 'gold standard' of evaluations for appraising cases.[45] The legal memorandum giving legal advice to officers quotes from the COT report that '[t]here was no evidence of mutagenicity, carcinogenicity or tetatogenicity. It was considered that the available data did not, in general, raise concerns regarding the affects [sic] of CS spray itself'.[46] In addressing why there had been so many reports in police magazines and elsewhere of CS being 'unsafe', the memo commented: 'There was concern expressed but the government COT report has addressed this. These magazine and newspaper reports are effectively out of date.'[47] As with the Home Office letter mentioned above, no attention is given in this memo to the 'sparsity' of data on CS and MIBK as a spray or the 1996 and 1997 Porton Down reports. In relation to the COT report, what the report actually meant and what 'facts' it substantiated are obviously matters interpreted in contrasting ways. No mention is made either in the legal advice of the call by COT to establish follow-up studies or of the lack of findings at the time the advice was given.[48] Following wider conventions, CS sprays are referred to as CS aerosols and evaluated in terms of CS as a gas.

There are further grounds that would make it difficult to assess officer compensation awards in civil courts (as opposed to workplace awards), even if figures about them were made publicly available. Officer court cases calling for compensation under £1000 ($1500) do not go to small-claims court. As officers are paid for their time off work, unless they are willing to bring a case for pain and injury, they are unlikely to seek civil redress in many cases. One of the former officers I spoke with who had suffered long-term ailments suggested that few of his past colleagues were likely to come forward out of fear for possible repercussions. Although this person knew of three officers in his force who had experienced lasting injury (skin complications, allergic reactions), none of them pursued compensation claims supposedly 'because they are young in service and afraid they will lose their jobs, like [David] Power'.

Flow My Tears the (Former) Policeman Said[49]

The case of David Power provides an illustration of how indeterminacies associated with the effects of CS sprays come to bear on a particular case. Power became a member of the Dyfed-Powys force in 1997. In the space of a few months he was exposed to CS spray four times, an exceptionally high rate: once in training, twice when other officers used the sprays in conflicts, and 'once' from his own use of the spray and his subsequent wearing of clothes contaminated in that incident. After the third and fourth exposures, Power experienced severe blistering and inflammation on one arm. Because of the injuries sustained, he was retired on medical grounds and received an injury pension and small damage award. Since then he has been attempting to receive an additional settlement in the civil courts for his loss of career and the injuries incurred. Although somewhat supported by his local Federation representative, owing to the types of legal advice given above, it has not financially supported his case. Power has raised concerns about his lasting injuries with numerous individuals in the Home Office and elsewhere over the last few years in an attempt to substantiate claims about the hazards of CS sprays. Figure 8.2 provides an excerpt from one such correspondence between his mother and a Home Office official.

The letter repeats, almost verbatim, many of the statements made above. As with the earlier Home Office statement, the COT report is drawn on in such a way as to support the safety and rigor of testing rather than drawing attention to the uncertainties mentioned. Here, the

---

FIGURE 8.2
ONE OF MANY ESSENTIALLY IDENTICAL BUREAUCRATIC LETTERS EMANATING
FROM THE HOME OFFICE REGARDING CS SPRAYS?

19 October 2001

Dear Mrs. Power

I refer to your letter of 18 July to the Home Secretary concerning your son and the use of CS spray ...

Firstly, may I express my sympathies to both you, and especially to your son, regarding this matter and can fully understand the sentiments expressed in your letter.

To begin with I will give you the historical background to the deployment of CS by the police in England and Wales. In the light of police trials of CS incapacitants in 1996, the Association of Chief Police Officers (ACPO) decided to recommend the adoption of hand held CS spray devices ...

The health effects of CS have been thoroughly researched, to a level similar to that which would be required for a pharmaceutical drug. This has established that CS presents no significant risk to human health. In September 1998, the Department of Health announced their decision to refer CS spray to the independent expert committees on Toxicity, Mutagenicity and Carcinogenicity, a decision which was welcomed by Home Office Ministers. In September 1999, the Committees published their report, which concluded that 'the available data did not, in general, raise concerns regarding the health effects of CS spray itself'. The report also acknowledged the vast number of toxicological tests that have been carried out on CS since it was first synthesised by stating that 'there are considerable data available to assess the toxicity of CS'. ACPO have given very careful consideration to how CS spray should be used, and to the aftercare of people who are sprayed with it, and have issued detailed guidelines to all forces in England and Wales.

I should add that the Home Secretary does not have to approve equipment for use by the police in public order situations. He provides advice on the equipment available, but the decision is ultimately for each Chief Constable. The decision to deploy CS in Dyfed Powys would be entirely a matter for the Chief Constable of that force ...

Yours sincerely,

Jonathan Batt
Action Against Crime and Disorder Unit

---

thing mainly in question is CS (implicitly as a gas or aerosol) rather than CS sprays. The last paragraph draws attention to the issue of responsibility for decisions. As stated, the final responsibility (and liability) rests with the Chief Constable of each force. The Home Office, through its various studies and testing procedures, only gives advice. This means that officers, like David Power, who seek compensation must take action against their Chief Constable – a move he has not relished. In any case, without any financial backing Power has so far not been able to bring a case to court.

Key issues of dispute regarding Power's injuries were whether or not he experienced a skin irritation or allergic reaction and whether that derived from the CS, MIBK, or the combination of the two. Another initial area of debate was whether the effects experienced could have been mitigated through different actions by the force and Power. As the Home Office further commented in the letter cited in Figure 8.2:

> PSDB add that decontamination of CS is a very important issue. If an officer has been in contact with CS spray and it is likely that particles have transferred to his clothing, it is very important that these clothes are decontaminated. Folding the clothes and putting them in the locker will not allow the particles a chance to be removed. Thorough ventilation and washing, where appropriate, should be undertaken every time CS particles are present.

Decontamination has been one of the major areas of general concern in relation to CS sprays. As part of Sussex's review of the sprays, in early 1998 Gregory and Knill identified several instances of long-term injury (allergic dermatitis, prolonged respiratory problems) to officers and others believed to derive from indirect CS spray exposure. Anecdotal injuries were identified by conversation with officers in other forces. In describing these, the authors found it necessary to add:

> It has been very difficult to get forces to describe problems they have had with contamination as they tend to want only to sell the advantages of CS and they maintain that whatever the disadvantages, CS is worth it. These people are usually the hands-on trainers. It is more difficult to get a hold of other contacts.[50]

Questions have been raised about just what steps are necessary and feasible to minimize hazards. 1996 ACPO guidelines for officers suggested that those sprayed 'should be removed to an uncontaminated area where they can be exposed to cool fresh air. This will permit the CS particles to be blown off the body.'[51] As part of what was typified as a wider misunderstanding between CS aerosols and sprays, Jones argued that while this suggestion might be appropriate for those exposed to CS aerosols, it is inappropriate for sprays because the latter leave significant amounts of residual chemicals deposited on the skin and clothing.[52] It was not until five years after the introduction of the sprays that the PSDB suggested that forces use sodium metabisulphite as a decontaminant for vehicles and buildings, though this had been identified publicly in 1991 for humans and buildings.[53] No decontaminant has so far been introduced for people. This not only raises concerns about unnecessary injury, but also about the duration over which recipients experience pain from exposure.

The introduction of even relatively simple devices, such as CS sprays, is not characterized by the simple insertion of a technology into a given setting. In relation to decontamination, there have been various calls about the need for accompanying socio-technical changes in the operating environment, such as providing adequate ventilation systems in sprayed areas or areas where sprayed individuals remain. In police vans or custody suites this might prove difficult. Noting this is not just to suggest that CS sprays require various conditions to function 'properly'. As previously argued, in cases where adverse effects take place from exposure to the sprays, the casual roles attributed to the 'context' or to the 'technology' are made in relation to each other.

Used? Misused? Abused?

To allay fears of their widespread use, the sprays were only supposed to be employed in specific ways for the defense of members of the police or the public in handling highly dangerous situations.[54] Although originally justified as a last-step measure for self-defense, a variety of reports have alleged that the sprays have been used in a wide range of circumstances and conditions.[55] Repeated, albeit anecdotal, reports have been made of the sprays being targeted against non-threatening individuals for 'offensive' subduing purposes rather than for officer or public protection.[56] The general thrust of such criticisms is that the sprays have become an easy option for officers where recourse to force is not the most appropriate means of handling a situation. The Police Complaints Authority (PCA) conducted a study of a limited sample of complaints for one year and found in 30 percent of cases that the sprays were said to be used under the one-meter minimum squirting distance; in 40 percent of cases that they were used in enclosed spaces where CS concentration levels can build up; and in 40 percent of cases it was determined that the sprays were not being used for self-defense. As a counter to critical claims about the potential for injury, the PCA found relatively temporary effects given in complaints. In only 6 percent of cases did individuals report 'serious injury', those being defined as injuries lasting longer than 2 to 3 hours and/or requiring hospital treatment.[57]

Senior officers have made adamant counter-responses to suggestions of the injudicious or ad hoc use of the sprays.[58] Much uncertainty and dispute stems from the relation of the technologies to other force options – whether, for instance, they are to be used before, after, or as a substitute for batons.[59] The potential transformative effects of the sprays have been acknowledged by some police forces. The risk assessment undertaken by the Sussex force noted the potential of sprays to dramatically alter how violent individuals are handled, perhaps discouraging officers from taking the time to talk individuals down.

These points are in addition to a variety of other officially acknowledged uncertainties related to the effects of the sprays that complicate recourse to them. In general, the inability of police officers to gauge the pain they are inflicting makes determining the appropriate application of the sprays somewhat problematic. Alternative characterizations of their effectiveness have been put forward. Officers have reported significant time delays (more than 5 seconds) in the sprays working in nearly a quarter of instances.[60] Partially due to the inadequacies of existing reporting procedures and partially for reasons that are unclear, in 20 percent of cases the sprays have been recorded as having no effect. The uncertainty over effectiveness thereby complicates making a quick and clear appraisal of their effects and of what subsequent action might be required.

A Tempest in a Teacup? A Storming Brewing?

How serious are the dangers associated with CS sprays? Where do the causes of concern stem from? What steps ought to be taken in the future to mitigate problems? As argued above, despite various studies undertaken, lingering questions and uncertainties can be posed about the sprays. Much of the debate turns on distinctions offered. So, for some, CS sprays are safe because CS is said to be safe, and CS sprays in Britain are safe because other such CS incapacitants are said to be relatively safe elsewhere. In a survey of counterclaims to the official position of government spokespersons, these similarities were said to be ill-founded or inappropriate. The positions taken on distinctions are highly consequential in the way claims about safety are built up over time.

As has been suggested, in interpreting the existing data, concerns about its credibility and comprehensiveness of claims come into play. As one 'fact' is stated, several additional questions can be posed. A significant effort has to be expended to address basic questions about safety. Does David Power's experience represent just one of a 'small number of cases [that] have now come to light which suggest that the CS spray has caused an allergic reaction in a small number of individuals', as suggested by the Home Office?[61] To what extent will more widespread severe reactions take place after multiple exposures to the sprays become more common? To what extent might officers not come forward about long-term ailments due to fears over the loss of their job?

How many officers and members of the public have already been affected and how severely? The failure of governing bodies to make the most basic information available presents major impediments, but it seems highly unlikely that such material in itself can resolve the disagreements made.

In overall evaluations, any negative effects suffered, though, would have to be seen in relation to claims about the positive benefits of deploying the sprays. As mentioned, one of the reasons sprays were introduced was to reduce assaults on officers. To what extent assaults (and therefore injuries) have been lowered is open to question. During the trials in 1996, officers perceived a marked reduction in assaults. However, force assault information gathered from trial locations did not suggest any greater reduction in assaults.[62] While the number of assaults against police in England and Wales fell from 15,500 in 1996/97 (the year of the wide-scale uptake of the sprays) to 12,500 in 1998/99,[63] they have been in general decline since 1991/2, when over 19,000 were recorded.[64]

There are, then, various uncertainties and disagreements associated with the operational deployment and health effects of the sprays. Important questions exist about who should be and has been targeted, as well as in what manner. Assessment of health effects and the proper operational condition of deployment are established in a reciprocal relation. What one believes about the adherence of the police to the rules affects the evaluation of the health dangers posed by the sprays 'themselves', as opposed to complications that might arise due to their 'misuse' or unadvised use. The reciprocity of the causes is source of much disagreement about blame. Any number of factors could be attributed with fault when complaints are made or there are serious reactions: the chemical composition of the sprays (whether that be CS, MIBK, or the combination of the two), their uncontrollable variability in effects across the population, the actions of officers (for instance, firing the sprays too close to the target), the dissemination characteristics of the jet sprays, the behavior of recipients, etc. Any assessment of where blame lies in relation to specific events will have to negotiate with such factors, typically making determinations well after the event.

At the general level of whether and when the CS sprays are appropriate use-of-force options, overall determinations of this are bound up with the same sorts of indeterminacy. For their part, organizations such as the PSDB and the Home Office have publicly focused on the importance of finding a new solvent for the current sprays and decontamination steps to reduce risks; though after six years of reportedly searching for an alternative, none has as yet been offered and no human decontaminant has been recommended.

As suggested in the last chapter, there is a fundamental problem in any sort of overall evaluation of the sprays, whether that might be positive or negative. General claims that give some policy or other practical guidance must be reconciled with the appropriateness of the sprays in specific settings. In this sort of situation, it was also suggested that a helpful analytical way of working with indeterminacies and conditionalities is to consider how they are distributed and the implications therein for notions of responsibility and legitimacy. It is to these issues that we now turn.

MANAGING IN THE MIRE

As suggested already, the management of disputes about CS sprays as it relates to the location and distribution of responsibility is a matter of some importance. Although a variety of organizations have contributed to the evaluation of the incapacitant, it is difficult to find any prepared to take responsibility for the deployment decision. The dynamics of attributing responsibility are almost completely opposite to those of building up facts about safety.[65]

Despite claims about CS or CS sprays being tested to a level of that required of a pharmaceutical drug, the sprays (or CS alone) have not been subject to formal approval by the Medicines Control Agency. The incapacitants eventually became justified as part of attempts to adhere to more general health and safety workforce standards (Police [Health and Safety] Regulations 1999 and the Police [Health and Safety] Act 1997). These standards require employers to minimize the reasonably foreseeable effects of workplace equipment that affect health and safety. Standard and well-worn regulatory procedures are in place for handling health and safety workplace guidelines, overseen by the Health and Safety Executive. In the case of police self-defense and restraint equipment, however, a special exemption was given. The safety of such devices has been primarily a matter for the UK Home Office, which did not have well-established procedures.

The Home Office and Ministers sought technical advice from the PSDB. Yet, this has not implied an official acknowledgement of the latter's responsibility. In relation to criticisms about their toxicological appraisal, a Home Office official explained:

> [The PSDB's] main role is to provide unbiased scientific and technical advice and guidance to their customers, which includes the Police Service. PSDB staff are mainly scientists and engineers and they do not, and have never claimed to

have, any toxicological expertise within the Branch and they always seek advice in this area from outside experts.[66]

As outlined above, much of that sub-contracted toxicological expertise came in the form of literature reviews and advice from Porton Down, though central conclusions made by the latter were not implemented.

COT and the other committees gave tentative and qualified conclusions, themselves predicated on assumptions about operational deployments. Although the Department of Health provided the secretariat to COT and has otherwise advised about the safety of solvents as mentioned above, correspondence with the author suggests the Department has not had 'any policy role in the approval of chemical incapacitants – this is for the Home Office and the Chief Police Officers'.[67] As mentioned previously, solicitors representing officers and others 'just' work with the conclusions reached by COT. The cumulative effect of these statements is to suggest that it is Chief Constables and they alone who are ultimately liable and even generally responsible for decisions. The division of labor and the scope that this brings for the displacement of responsibility in the overall regulation of the CS sprays, as outlined here, arguably share many of the characteristics of the distribution of responsibility for other non-lethals discussed throughout Chapter 5.

To focus on this broad level alone, though, limits the insights of analysis. Chapter 6 argued that different interpretations of what technology is and what it promises to do are formed in relation to an understanding of rules governing their use. In the case of the CS sprays, various rules were given about how they should be used, including from what distance they should be fired, for what duration, in what situations, and with what sort of aftercare procedures. Disputes about what 'the effects' of non-lethal weapons are depend on assessments of the feasibility and likelihood of adhering to such guidelines.

Arguably, safety concerns vis-à-vis the assessment of the sprays were not the only major factors bearing on their uptake. While the sprays might be only one option in the arsenal of US and other police forces, in the British context their deployment, for some, threatened to compromise the distinctive basis of legitimacy for the police. Whether in substance or in myth only, the view is often put forward that a special reverence exists for the police in mainland Britain;[68] particularly due to a lack of routinely armed officers, the British 'bobby' is supposed to be non-threatening. The relatively vulnerable position of 'bobbies' requires a relation to the public based on mutual respect. This sort of policing is supposed to impart a level of legitimacy to the British police that means the recourse to coercive measures is less called for than in the United

States or continental Europe. Waddington and Hamilton[69] argue that the emphasis placed on the 'special relationship' with the public both under-pins, and yet poses a constant threat to, the basis of public support for the police. Changes in the operation of the police are scrutinized as to whether they are consistent with 'traditional' forms of policing, not least by the police themselves. In the case of the sprays, the uptake of the sprays represented, for some, a further and unproductive 'tooling up' of the British police that would ultimately prove counterproductive.[70]

Following the situation as I have presented it above, these general considerations, combined with the controversy surrounding the sprays, meant that there were acute tensions regarding their internal and exter-nal legitimatization through the devising of guidelines. If the rules surrounding the sprays were highly proscriptive, then, they would lack flexibility, act as a method for post-hoc assessments and no doubt appear quite unrealistic to many officers. As argued in Chapter 6, discretion is a key part of policing and one that poses a number of dilemmas. Rules about the use of force, no matter how elaborate or well-communicated, would never be exhaustive of how force ought to be used in all situations. However, perhaps particularly in the British case with regard to 'policing by consent', rules without any teeth would lack credibility for the public as well as the police and risk allegations of an ad hoc and uncontrolled deployment.

In 1996, ACPO produced guidelines to advise chief constables and individual officers about the appropriate use of the sprays. Although individual police forces were able to revise the ACPO guidelines, few, if any, have done so.[71] The 1996 Guidelines 'resolve' the tension above by deferring the moment of its resolution. The decision to use the spray is treated as an individual one for officers who will be accountable to the law. As such, ACPO reinforced the legal and occupational discretion of individual officers and the policies of police-chiefs. The non-legally-binding 1996 Guidelines specify that the incapacitants should be used against:

(1) those offering a level of violence which cannot be appro-priately dealt with by 'empty hands' techniques, and

(2) violent offenders, other than those armed with firearms or similar remote injury weapons, where failure to induce 'immediate' incapacitation would increase the risks to all present.[72]

Several proscriptions are made. The use of the spray against those with a firearm, for instance, is deemed 'inappropriate and should not be considered'. Officers should not use the sprays within three feet of a target 'unless life is at risk'. Despite the ultimately provisional character of such stipulations, their 'inflexibility' brought a rewriting in 1999.[73]

As opposed to the 1996 *Guidelines*, the 1999 policy is one of *Guidance*. The guidance is meant to 'support and inform' individual officers in determining what constitutes reasonable and necessary force. The major point of departure is in the appropriate contexts for acceptability. The 1999 advice states that the spray is a tactical option 'available to an officer who is faced with violence or the *threat* of violence [my emphasis]'.[74] The provisional character of the proscriptions given to officers is reinforced by the latest advice. So, the guidance advises officers not to use the sprays below three feet, unless it cannot be avoided. Possibilities for cross-contamination means the sprays should not be used in crowd-control situations unless the officer can justify his or her actions. The decision of whether to spray the mentally ill is one where an officer may consult others, but the final decision rests with the officer. The use of the incapacitant against those armed with firearms or those in control of motor vehicles needs to be 'carefully considered'. Cautionary issues are noted, but only noted.

To different degrees, the 1996 ACPO Guidelines and 1999 Guidance handle the indeterminacies and uncertainties surrounding the sprays and the desire for internal and external legitimacy by devolving responsibility for appropriate use down to individual officers. The provisions in place lay out general policy directives while acknowledging the central role of discretion in policing. Where proscriptions are made, they could hardly be said to be definitive. The general policy is that it is for individual officers to decide on and justify the appropriateness of the sprays on a case-by-case basis. Conceived as such, the merits of the sprays depend on the intent and actions of their users.

Whether intended or not, this focus on the practices of the direct users deflects attention away from critical questions about government and corporate responsibility regarding safety concerns, as well as the role of senior police management in ensuring a proper operational deployment. While there are risks acknowledged with sprays, these are portrayed as manageable and established ones associated with the use of any such chemical incapacitant. In this regard, though there exists the possibility of severe but unforeseeable reactions, the sprays are safe, being tested to the level of a pharmaceutical drug. The quote from the COT report that '*available* data did not in general raise concerns regarding the health risks of CS spray itself' has been cited (typically without italicizing 'available' and without further qualification) in a variety of statements – Home Office press reports, Home Office correspondence, Police Complaints Authority evaluations, legal advice, etc. –

as a way of limiting debates about health risks. In relation to questions about the solvent, a major government review of non-lethal weapon options in 2001 for Northern Ireland and mainland Britain stated: 'There have been one or two reports suggesting that MIBK is not the ideal solution, and that it might carry risks that another solvent might not.' The truncations associated with these types of statement indicate how declarations of safety are acknowledged.

With the management of many of the health considerations and presumptions in official circles against widespread abuse of the sprays, the main focus of attention becomes the instructions given to officers. Problems with the sprays become defined in terms of the inadequacies of existing training provisions. The focus on training is also shared by quasi-regulatory bodies such as the PCA. Its overwhelming prescriptive emphasis has been on training and guidance to officers (for instance, in relation to the distance at which people are sprayed).[75] With the attention on training provisions, the central question for criticism becomes one of asking whether officers are using the sprays properly or improperly. Critiques of the sprays thus become critiques of the actions of officers (as users or trainers). This framing is no doubt part of the reason why police bodies have responded so adamantly against critiques of 'the sprays'. Officers are said to do a difficult job in sometimes violent situations and need the best protection.[76] When this justification is married with the importance of discretion, the possibilities for reprimanding officers is limited. Cases of severe reaction, for instance, might be attributed to officers holding the sprays too close to the targets, rather than being an effect of the composition of the sprays. While the former is regrettable, its importance is mitigated because of the volatility of the circumstances of use. Although senior officers might defer the indeterminacies and disagreements surrounding the sprays down to rank-and-file ones, given the importance of officer discretion and the difficulties of substantiating complaints about police use-of-force, the potential for the diffusion of responsibility at this level is more than a mere possibility.[77]

Given the considerations above, debates about the sprays have tended to degrade into sets of oppositions: side effects vs officer protection, safety vs liberty, use vs abuse, safe vs unsafe, officers vs criminals, etc. Proponents and critics typically take opposite sides in this debate and both appeal to definite accounts of technology to support their positions. Civil rights groups such as Liberty, human rights organizations such as Amnesty International and anti-racist groups such as the Newham Monitoring Project have called for the deployment of the sprays to be suspended until the facts regarding their safety and operational deployment are known. Certainly, while there are pragmatic reasons for seeking a clearer understanding of the 'facts of the matter', as suggested, there are major problems in cutting through assemblages of humans and non-humans in order to attribute definitive effects to technology or actions to individuals. Not least is the possibility for alternative assessments between interested parties about what constitutes reasonable force.

A different critical response to some of the points noted above has been to seek to define generalizable and definite prescriptions about how the sprays ought to be deployed. Concerned Parliamentary representatives have suggested that CS spray should not be used against children or those known to have mental-health problems. Such suggestions have been criticized for being unrealistic in practice (how young is 'too' young?) and not allowing officers to independently deal with particular situations. Young people or the mentally ill may be extremely violent and officers are thus said to need to defend themselves.[78] The PCA recommends that officers not use CS sprays against those 'seen to be' displaying 'acute behavioural disturbance', due to risks of sudden death during restraint for such individuals.[79] Whatever the merits of such proscriptions, the claims and counterclaims put forward for them represent attempts to occupy a pole in the general/particular spectrum.

This section has argued that in the case of the CS sprays, discussions about the capacities of technology have been polarized as different but definitive accounts of technology are put forward. It has further been suggested that in the search for definitive claims, the manner in which discussions divide into opposing camps means that the merits of the sprays are framed within the fairly narrow terms of the use, misuse and abuse of technology. As such, the complexity of the debate about the sprays and the constitutive functions of the socio-technical processes at work are lost in favor of a discussion of the 'actual' health effects of the sprays and the intent of officers.

These dynamics could hardly be said to be unique to the case of CS sprays. As argued in previous chapters, the use/abuse model is prevalent in claims by critics and proponents alike for all sorts of non-lethal weapons. In this model, technology is evaluated as good or bad, appropriate or inappropriate, depending on its usage. In other words, the object in question is taken as essentially neutral, this despite the recognition of the negotiated status of technology and uncertainties associated with it. While such a framing is perhaps pervasive in accounts of technology, it ill-serves attempts to devise processes for learning from the complex experiences in conditions of disagreement and uncertainty.

As argued here, commenting on what non-lethals do or do not do is not simply a matter of specifying their actual effects or merely noting the scope for multiple assessments, but is instead the building of an approach sensitive to the conditions under which interpretations are made and how the legitimacy of claims are secured. In the most robust, honest and demanding form of such an approach, the importance of these issues would extend to an examination of the claims of the analyst. This chapter, and the book as a whole, has wrestled throughout with the problem of how to speak about the characteristics and effects of technology as both 'real' and 'discernable' matters deserving attention and 'claimed' attributions whose importance is highly negotiable. In focusing on the management of the indeterminacies, disagreements and uncertainties, this analysis has displaced rather than superseded the thorny question about how to specify the characteristics of technology.[80] In many ways, this displacement activity is a matter for you, the reader, to resolve. Furthermore, the conceptual propositions made can be analyzed in terms of the general-particular tension noted previously. An attempt has been made in this chapter to offer an argument that is seen as legitimate because it is at once both relevant to the case at hand, but also informs wider discussion about discretion, responsibility and technology.[81] To the extent that this has taken place, the outcome is a contingent achievement.

A SPRAY TOO FAR?

This chapter has surveyed claims and counterclaims about the safety of CS sprays in Britain. Various grounds have been given for casting misgivings on the highly optimistic statements about the safety of these devices. The survey of the issues at stake has been undertaken with the intent to illustrate something of the indeterminacies and uncertainties about non-lethal weapons. In this complex, disputed and disputable case, what can be said is that whatever the 'facts of the matter' are, they are not the basis for decisions. Debates about the sprays are debates about what counts as equivalent to what, what counts as valid evidence and what inferences can be made.

Beyond that, however, this chapter has sought to understand the attribution of responsibility as a product of particular organizational strategies. As argued here, the prevalence of framing about the use, abuse, or misuse can be understood, at least in part, as a product of attempts to reconcile particular and general claims. When one pays attention to the management of indeterminacy and uncertainty and its deferral, it is possible to see the contingency of such a framing; the effort behind attributing blame to downstream actors. At stake in discussions about uncertainty and indeterminacy is not just how these are distributed, but how such a distribution in turn helps to constitute an understanding of technology, users and notions of responsibility.

Whereas many analyses stress the importance of training as a *factor* that determines the legitimacy of non-lethals, that focus has been treated here as an *outcome* of the way in which indeterminacy and uncertainty are distributed.

NOTES

1. The evidence for this chapter is based on published medical literature, government-related research, journalist accounts as well as 35 interviews conducted during 2000–1 with the Home Office, government agencies, police officers, Members of Parliament, research scientists, lawyers and others associated with the sprays. The use of CS sprays is considered only as it relates to England and Wales, since they are not deployed in Northern Ireland, and Scottish forces have generally been slower and more reluctant to take up the sprays; see Tyler, L. and King, L., 'Arming a Traditionally Disarmed Police', *Policing*, 23, 3 (2000), 390–400.
2. For a discussion of this see Buttle, J., *The Influence of the United States on the British Force Regarding the Use of Chemical Incapacitant Sprays* (MA in Comparative Criminology & Criminal Justice, Bangor: University of Wales, 1999).
3. Association of Chief Police Officers, *Police Trials of CS for Self-Defence in England and Wales – 1996, ACPO Guidelines for Use* (London: ACPO, 1996) and Kock, E. and Rix, B., *A Review of Police Trials of the CS Aerosol Incapacitant* (London: Home Office Police Research Group, 1996).
4. See e.g. Campbell, D., 'Police Test CS Sprays to Combat Violence', *The Guardian*, 19 January 1996, 5.
5. As in 'Is it Safe?' *Police Review*, 20 November 1998, 4.
6. See Trevisick, S., *Dispatches* (London: Channel 4 and Liberty Publication, 1996); Schmid, E. and Bauchinger, M., 'Analysis of the Aneuploidy Inducing Capacity of 2-Chlorobenzylidene Malonitrile (CS) and Metabolites in V79 Chinese Hamster Class', *Mutagenesis*, 4 (1991), 303–5; Hu, H., Fine J., Epstein, P., Kelsey, K., Reynolds, P. and Walker, B., 'Tear Gas – Harassing Agent or Toxic Chemical Weapon?, *JAMA*, 262 (1995), 660–3; Ballantyne, B., Gall, D. and Robson, D., 'Effects on Man of Drenching with Dilute Solutions of CS and CR', *Medicine, Science and Law*, 16 (1976), 159–70; and Jason-Lloyd, L., 'CS Gas', *New Law Journal* (1991), 1043–5.
7. See Michael, A., 'CS Gas', *Hansard*, 16 June (London: HMSO, 1998): column 173 in relation to the arguments presented here.
8. Williams, Lord, Statement on CS sprays: Police use, *House of Lords Hansard*, 27 July (London: HMSO, 1998), column 1190.
9. British Broadcasting Corporation (BBC) News, 'Life-Saver or Deadly Tool?', *BBC News*, 8 July 1999.
10. For a wider discussion of this point see Gregory, P. and Knill, L., *CS Incapacitant* (Lewes: Sussex Police, 1998).
11. Wadham, J., Preface in *Dispatches* (London: Channel 4 and Liberty Publication, 1996).
12. Lancet ' "Safety" of Chemical Batons', *Lancet*, 352, 9123 (1998), 159.
13. While Nottinghamshire later introduced CS sprays when a new chief constable was appointed, Sussex and Northamptonshire have still refused at the time of writing.
14. Boateng, P., *Home Office Communication to Matthew Taylor MP*, PRO/98 9/11/4, 8 January 1998.
15. Jowell, T., *Department of Health Communication to Matthew Taylor MP*, POH(3)5592/141, 30 March 1999.
16. This quote is the motto of the Chemical and Biological Defence Establishment.
17. Gray, P., 'CS Gas is not a Chemical Means of Restraining a Person', *British Medical Journal*, 314 (1997), 1353 and Gray, P., 'Formulation Affects Toxicity', *British Medical Journal*, 321, 1 July (2000), 46.
18. As in Michael, 'CS Gas', *Hansard*; Fraunfelder, F., 'Is CS Gas Dangerous?', *British Medical*

Journal, 320 (2000), 458–9 and Yih, J., 'CS Gas Injury to the Eye', *British Medical Journal*, 311 (2001), 276.

19. Jones, R., 'Are CS Sprays Safe?', *Lancet*, 350 (1997), 606.
20. Trevisick, *Dispatches*.
21. Parneix-Spake, A., Theisen, A., Roujeau, J. and Revuz, J., 'Severe Cutaneous Reactions to Self Defence Sprays', *Arch Dermatol*, 129 (1993), 913.
22. Rice, P., Dyson, E. and Upshall, D., *A Review of the Toxicology of Methyl Iosbutyl Ketone and Methylene Chloride* (Salisbury: Chemical and Biological Defence Establishment, 1996).
23. Quoted from West Midlands Health and Safety Advice Centre (WMHSC), *CS Incapacitant Spray* (Birmingham: West Midlands Health and Safety Advice Centre, 1996), 3.
24. Rice, P., Jones, D. and Stanton, D., *A Literature Review of the Solvents Suitable for the Police CS Spray Device* (Salisbury: Chemical and Biological Defence Establishment, 1997), 7.
25. Several people made 'off-the-record' comments to me.
26. See Jenkins, C., 'Safety of CS Spray Still Undecided as Review Committee Stalls Again', *Police Review*, 30 July 1999, 4.
27. Ibid.
28. Committees on Toxicity, Mutagenicity, and Carcinogenicity of Chemicals in Food, Consumer Products, and the Environment, *Statement on 2-Chlorobenzylidene Malononitrile (CS) and CS Spray* (London: HMSO, 1999), 11.
29. Benford, D., Personal Interview, 8 January 2001.
30. Kammen, D. and Hassenzahl, D., *Should We Risk It?* (Princeton, NJ: Princeton University Press, 1999); Richardson, G., Ogus, A. and Burrows, P., *Policing Pollution* (Oxford: Clarendon, 1982); and Thornton, J., *Pandora's Poison* (London: MIT Press, 2000).
31. Blowe, K., *National Campaign Against CS Spray* (London: NMP, 2000).
32. Home Office, *Home Office Welcomes Independent Scrutiny Findings on CS Spray*, Press release, 294/99, 23 September 1999.
33. Motto of the UK Home Office.
34. See Patton, L., 'CS Gas "Guinea Pig" Complains of Eye Damage', *The Guardian*, 29 January 1998, 4.
35. Gregory and Knill, *CS Incapacitant*, 10.
36. Gray, 'CS Gas is Not a Chemical Means of Restraining a Person', and Gray, 'Formulation Affects Toxicity'.
37. Worthington , E. and Nee, P., 'CS Exposure' *Journal of Accident and Emergency Medicine*, 16 (1999), 168–70.
38. Jenkins, C., 'Pain Reaction', *Police Review*, 1 October 1999, 18.
39. E.g. Hindley, C., 'Cops in CS Storm', *Manchester Evening News*, 18 September 2001; Flury, A., 'Pregnant Women Sprayed with CS Gas May Sue', *The Times*, 15 June 1998; and Stern, C., 'Detective has Heart Attack Training with CS Gas Spray', *The Mail on Sunday*, 25 October 1998.
40. E.g. Brooke, C., 'Father Dies after Police Use of CS Gas', *Daily Mail*, 6 November 1998; Stratford and Newham Express, 'CS Gas Death', *Stratford and Newham Express*, 6 September 1997; and Dobson, R., 'Man's Death Raises Alarm', *Independent*, 11 August 1998.
41. See Wheeler, H., MacLehose, R., Euripidou, E. and Murray, V., 'Surveillance into Crowd Control Agents', *The Lancet*, 352 (1998), 991–2.
42. Here I refer to injuries consistent with those observed by Ro, Y. and Lee, C., 'Tear Gas Dermatitis', *International Journal of Dermatology*, 30, 8 (1991), 576–7. This was conveyed in personal interview with Euripides Euripidou on 14 November 2000.
43. Boateng, P., 'CS Sprays', *House of Commons Hansard*, 27 January (London: HMSO, 1998), column 449.
44. Harrison, John, Personal Correspondence, 2 September 2000.
45. Care, A., Personal Interview, 30 August 2001.
46. Care, A., *Memorandum*, December (London: Russell, Jones & Walker, 2000), 2.
47. Ibid.
48. Correspondence between the Home Office and the author indicated that a follow-up was under way and that the results would be released in late 2002.

49. Title taken from Philip K. Dick's book *Flow My Tears, the Policeman Said*, a novel involving multiple realities in a police state.
50. Gregory and Knill, *CS Incapacitant*, 17.
51. Association of Chief Police Officers, *Police Trials of CS for Self-Defence in England and Wales – 1996 Guidelines for Use* (London: ACPO, 1996), 6.
52. Jones, 'Are CS Sprays Safe?', *Lancet*, 606.
53. Jones, R., 'CS Gas', *Military Medicine*, 156, 11 (1991), A6–7.
54. See Steele, J., '2,500 Policemen to be Issued with CS Sprays Today', *Electronic Telegraph*, 26 June 1998. See www.telegraph.co.uk
55. See e.g. Bell, F. and Thomas, B., 'Police Use of CS Spray', *Mental Health Care* 1 (12) August 1988, 402–4; Boycott, O., 'Guidelines for CS Gas "Ignored"', *The Guardian*, 14 May 1996, 7; Brindle, D., 'Police Asked to Justify Use of CS Gas Spray on Boy, 14', *The Guardian*, 8 January 1998, 10; and Campbell, D., 'Police Defend their Use of CS Spray on Mother', *The Guardian*, 28 August 1996, 4.
56. Johnston, P., 'Police "Using CS Sprays Too Often"', *Electronic Telegraph*, 26 June 1998. See www.telegraph.co.uk
57. Police Complaints Authority [PCA], *CS Spray*, March (London: Police Complaints Authority, 2000).
58. See Elliot, M., 'PCA Claims have Damaged the Reputation of CS Spray', *Police Review*, 30 July 1999 and Jenkins, C., 'South East JBBs Criticise CS Report', *Police Review*, 16 July 1999.
59. Compare for instance Graves, D. and Johnston, P., 'Police Tests on CS Gas to Go Ahead', *Electronic Telegraph*, 4 January 1996 and Steele, J., 'Fears Linger as Police Take CS Gas on the Beat', *Electronic Telegraph*, 26 June 1998.
60. Kock and Rix, *A Review of Police Trials of the CS Aerosol Incapacitant*.
61. Batt, J., *Letter to Mrs. Power – Home Office Communication*, 19 October 2001.
62. Kock and Rix, *A Review of Police Trials of the CS Aerosol Incapacitant*.
63. Police Complaints Authority [PCA], *CS Spray*.
64. For a more in-depth consideration of assaults see Rappert, B., 'Constructions of Legitimate Force', *British Journal of Criminology*, 42, 4 (2002).
65. For a general discussion of these points see Latour, B., *Science in Action* (Milton Keynes: Open University Press, 1987).
66. Batt, J., *Letter to Mrs. Power – Home Office Communication*, 4 March 2002.
67. Benford, D., Personal Correspondence with the author, 7 September 2001.
68. For a discussion of this issue see Reiner, R., *The Politics of the Police* (London: Harvester Weatsheaf, 1992).
69. Waddington, P. A. J. and Hamilton, M., 'The Impotence of the Powerful', *Sociology*, 31, 1, (1997), 91–109.
70. Ballantyne, R., 'It'll All End in Tears', *The Guardian*, 11 January 1996, 8.
71. See Police Complaints Authority, *CS Spray*.
72. Association of Chief Police Officers, *Police Trials of CS for Self-Defence in England and Wales – 1996*.
73. Neil Haynes, Association of Chief Police Officers, personal interview, 13 June 2000.
74. Association of Chief Police Officers, *CS Incapacitant Spray* (ACPO: London, 1999).
75. See Police Complaints Authority, *CS Spray*.
76. E.g. Straw, J., 'Policing (London)', *Hansard*, 16 July (London: HMSO, 1999), column 691.
77. The 2000 Police Complaints Authority report discusses some of the problems in proving allegations against the police of excessive force. See also Smith, G., 'Police Crime', Presented at State Crime and Corporate Violence Conference, 25–27 April 2000 (Bangor: University of Wales, 2000).
78. For a debate made with this frame of reference, see Hansard, 'CS Gas', *Hansard*, 12 May (London: HMSO, 2000), columns 1185–90.
79. Police Complaints Authority, *Policing Acute Behavioural Disturbance* (London: PCA, 2002).
80. My thanks to Steve Woolgar who brought up this set of issues.
81. For a further discussion see Lee, N., 'The Challenge of Childhood Distributions of Childhood's Ambiguity in Adult Institutions', *Childhood*, 6, 1 (1999), 455–74.

9

Gauging Electroshock Weapons

In the example of CS sprays considered in the last chapter, as well as in previous chapters, major government and non-governmental evaluations have been central pieces of evidence. Such evaluations, both in relation to one another and even within themselves, often provide alternative ways of making sense of technology and human agency in use-of-force situations. While it is apparent that evaluation studies can be drawn on as resources to selectively support positive or negative appraisals of technology, such an approach is arguably limited in furthering an understanding of the dynamics of disputes over legitimacy. As has been suggested throughout this book, in analyses of the non-lethal weapons (including the present analysis), pressing questions need to be asked about how, when and why particular descriptions are offered and what makes for a convincing line of argument. In Chapter 7, I suggested that the focus on the intent of users of CS sprays was a contingent outcome of the way debates were managed.

This chapter continues to examine questions about the legitimacy of non-lethal weapons and the function of key evaluations. It does so by considering the claims and counterclaims made about a class of non-lethals and a major trial of them. The class of weapon is electroshock and the trial is the introduction, in the mid-1990s, of stun guns and pepper sprays to all detention officers in Maricopa County, Arizona jails.[1]

This chapter provides something of an evaluation of the various evaluations made of the trial; 'something' in the sense that it is not the intent to adjudicate about what really happened in Maricopa County, as if this could simply be determined by a thorough reading of various evaluations. Rather, the purpose here is to examine this case with a view to asking what it suggests for the analysis of non-lethals, claims for the basis of their legitimacy and the evidential basis for claims. A telling of the various stories given about what happened in Maricopa County raises significant questions about what went wrong and how to ensure this does not happen again in the future.

Chapter 3 depicted a variety of devices that employ electrical currents for stated non-lethal ends. Some police officers and manufacturers are highly supportive of electroshock technology. For instance, Sid Heal has referred to electroshock weapons as 'magic bullets', because they match the public's expectation for weapons that are instantaneously effective, incur little or no physical harm, and cause reversible effects.[2] In short, for Heal, electricity-based weapons are the nearest things to the Star Trek phaser. Among manufacturers of electroshock, TASER producers have been some of the most vocal advocates. As the president of AIR TASER© (now called TASER International©) stated:

> It is unfortunate that our society [the United States] needs any weapons. But the fact remains that violence, like cancer, will continue to occur. And while chemotherapy is a highly unpleasant process, it is superior to the alternative of certain death. Our society has a cancer called gun violence, and non-lethal weapons can serve as the chemotherapy ... With an AIR TASER or other non-lethal weapon, no one dies. No one is crippled. No one is maimed. Medical costs are zero. There is no pain, no suffering.[3]

Smith claims that no deaths and no significant injuries have been reported about his company's TASER technology.[4] On the basis of key publications of operational experiences in the late 1980s, such as Ordog, Wasserberger, Schlater and Balasubramanium[5] and Kornblum and Reddy,[6] the non-lethality of TASER technology is 'now a fact'.[7]

In countries such as the United States, the lower-wattage versions are readily available through weapon shops or on-line sites. For instance, the 'Stun Monster' is claimed to be the strongest-hitting stun gun in the world by putting out 625,000 volts, yet 'it is totally non-lethal and will cause no permanent damage'.[8] For security forces open to liability claims for injuries they cause, the choice of *not* to adopt electrical devices for some is unacceptable. As former chief commissioner of the Victoria (Australia) Police Kel Glare (whose company is reported to have distribution rights for the AIR TASER in Australia) said, 'If there is now someone killed in circumstances where a Taser could have solved the problem without being lethal, police are vulnerable to a law suit'.[9] Assessments of TASERs as effective and benign are echoed halfway around the world. Sergeant Darren Laur of the Canadian Victoria Police Department states that TASER technology is 'well over studied' by the medical community.[10] He goes on to say:

I can not emphasise enough, that TASER pulse wave technology weapons that use 50,000 volts and 5 watts have been
medically proven to be safe when used on normal healthy
subjects. Although there are always risks using any force
option to control violent behavior, the medical risks posed by
the TASER are very minimal when compared to blunt trauma
injuries caused by empty hand impact techniques and baton
strikes, or even trauma caused by an officer's firearm.

Laur reports on trials where fourteen suicidal or violent individuals
were successfully subdued by use of TASERs. Moreover, in many of
these cases, merely displaying the device and making a verbal warning
of its shock potential was said to be enough to gain compliance. As in
other statements about the use of force, the basis of the credence of
these claims derives from the professional expertise and practical experience of officers in relation to the situations under question.[11]

Interfering Signals

In contrast to the images promoted in the company brochures and officers' reports cited above, a number of individuals have called into
question claims about the merits of electroshock devices. Here, as elsewhere with weapons designed to harm but not kill, there are questions
about the ability to discriminate effects. A consideration of such critical
points in the construction of facts brings up questions about the credibility of the sources, the significance attached to particular pieces of
evidence and the criteria for evaluation. In a study of the cardiac effects
of prominent stun-gun models on anaesthetized pigs, Roy and
Podgorski[12] found that such devices could cause ventricular fibrillation
if they were applied directly to the heart, and pump failure when
applied directly to the chest. For people utilizing pacemakers the
dangers were said to be particularly acute.

The Los Angeles Police Department was one of the first to take up
devices such as TASERs. If the conclusions of such experimental tests
are correct, then there are likely to be a number of deaths attributed to
such devices. And yet, the cause of death in police–citizen encounters
with electroshock technology, much as with pepper sprays, is open to
interpretation and negotiation. Field-based assessments by Ordog,
Wasserberger, Schlater and Balasubramanium[13] and Kornblum and
Reddy[14] of TASERs in Los Angeles have maintained that none of the
fatalities deriving from incidents involving the use of TASER weapons
were caused by the TASERs themselves.

Based on experience as a forensic pathologist in Los Angeles, Allen in turn has disputed such claims, contending that the conclusions were incomplete and flawed.[15] Although many (but not all) of the sixteen deaths examined by authors such as Kornblum and Reddy were associated with drug use, Allen argued that it was quite likely that this condition, intoxication, or others, such as heart disease, increased the lethality of TASERs. Key details that would need to be given in medical assessment (which were not presented by Kornblum and Reddy) include the location of the shot, the levels of drug concentrations and the time between electrical exposure and ventricular fibrillation. More insidiously, specifically as a rebuttal to the finding of Kornblum and Reddy, Allen contended that the TASER-related deaths in places such as Los Angeles would be higher were it not for certain practices:

> I was apparently one of only two medical examiners in the Los Angeles office to list taser on a death certificate. This was because pathologists in Los Angeles were under pressure from law-enforcement agencies to exclude the taser as a cause of death. [One] autopsy was performed in the presence of six upper-level law-enforcement agents who were confrontational and argumentative in their attempts to persuade me that death was caused by drowning in a few inches of water. I was not allowed to attend the death scene. I insisted that the cause of death would not be determined until all tests were complete. My opinion was widely and prematurely misquoted by the officers. Likewise, I was called into Dr. Kornblum's office to defend my investigation in something more akin to a disciplinary hearing than a scientific conference. In the end, Dr. Kornblum seemed to agree that the tasering was the immediate cause of death. Yet, in his article it is stated that 'the death clearly fits into cocaine category'.[16]

Such accounts of the conditions under which determinations of the causes of death are made stand in sharp contrast to definitive statements made elsewhere. Taking the allegations seriously would raise a host of issues. For instance, given initial evaluations early in the deployment of TASERs that refuted their causal role in deaths, might subsequent pathologists have felt it difficult to overturn such determinations? The Allen response also raises questions about the possible contributory, rather than causal, role of non-lethal weapons in fatalities. Various reports of injury and death have been associated with electroshock technologies, each raising questions about causality.[17]

Compounding such disputes about the causes of death in particular situations is the lack of significant regulatory or licencing procedures for

electroshock devices. TASERs are the only electroshock-related technology specifically approved as relatively safe by US government
agencies. In 1976, the US Consumer Product Safety Commission evaluated a 5-watt version of the TASER and found it relatively non-lethal
for 'normal and healthy persons'. O'Brien contended that the
Commission's findings were based on theoretical models that took as
their basis for evaluation the 'risk of unreasonable injury' rather than
the 'unreasonable risk of injury'.[18] According to him, stun-gun manufacturers have improperly drawn on these original findings to
substantiate the safety of other products. In 1991, he pointed to a lack
of national figures in the United States on deaths attributed to electrical
weapons, the limited research on effects, and various animal tests that
revealed significant dangers. This lack of a regulatory framework still
exists and is paralleled in Europe, where the use of such technology has
been much less widespread. European Commission agencies do test
electroshock equipment through the so-called 'CE' quality-approval
controls for electrical goods, though this safety marking applies to the
users of, rather than to *those targets or victims shocked by*, such
weapons.[19] Knowledge of the effects of electricity in the human body is
constantly developing. The possibility that electrical shocks may cause
motor neuron disease was suggested in 2001;[20] a link that could be
tested with respect to electroshock equipment through medium-term
surveillance procedures for monitoring those shocked.[21]

Adding to the qualifications and complications noted above are variations in effects due to recipients' characteristics (body temperature,
amount of clothing and skin moisture), the contact duration and the
areas targeted (the chest, eyes), and differences within and between
types of devices in terms of their power sources, peak voltage and electrical outputs. Once electroshock devices are used in connection with
other types of force, the possibilities for specifying likely effects of
weapons becomes even more problematic.

For those wishing to evaluate this technology, noting these considerations does not lead to clear evaluation implications. The factors noted
above obviously complicate statements made that electroshock technology poses no concerns. However, they also undermine generic criticisms
of this technology. If every type of electroshock weapon differs in its
characteristics, and its effects are conditional on situational factors, then
it becomes difficult to argue about the general unacceptability or risks
associated with this technology. Any definitive statement about the
'actual' effects of such a class of technology rests on a willingness to
suppress certain variables in order to make generalized claims.

The lack of research on electroshock technology and the contingencies and qualifications noted above make definitive safety assessments

rather problematic. This is illustrated in a report commissioned by TASER International in preparation for the trialing of its recently developed ADVANCED TASER device in a British police force.[22] As there is little in the way of clinical experience with this particular model, the hospital consultants commissioned conducted a medical literature review of the studies of electrical injury. Various controversial aspects of electroshock technology were raised as potential concerns but then downgraded because of inputs made via the manufacturer. So, the potential for heightened risks for those with pacemakers was called into question during the writing-up of the review by personal communication with the TASER International consultant, Dr. Stratbucker.[23] The review stated that Stratbucker reported not being able to duplicate the research results suggesting this and therefore that such studies were not credible. No further details are given as to the evidence offered. The consultants stated that they were not able to determine the 'debate' one way or the other, and they therefore concluded that the risks to those with pacemakers and defibrillators were 'quite small'. Similarly, the variability of effects due to alternative skin resistance and the subsequent implications for vital body organs was noted but downplayed because of a personal communication by Dr. Stratbucker during the drafting stage of the report. Stratbucker claimed that the high-frequency currents of Advanced TASER meant that they stayed near the body surface and did not affect vital organs. One of the authors of the review stated that he did not know how to treat such competing claims because he was not an expert on the specific subject.[24] The asymmetrical procedures regarding divergent knowledge claims reveal the types of contingency that influence the findings of such medical reviews of safety.

As well, because the review was concerned with evaluating 'objective' medical evidence, the organizational and institutional factors implicated in assessments of electroshock technology were not elaborated. This has important implications for the conclusions derived. So, while noting the medical claims presented by Allen and O'Brien, no mention is made of the wider questions which they raise about the determinants behind what evidence exists. In relation to possible fatalities, the review simply concludes that 'there is no convincing evidence directly implicating Taser weaponry in deaths of subjects in over 25 years' experience in America'.[25] Some cautionary points are noted, such as potential for greater harm for elderly subjects or those with pre-existing heart disease.[26]

For users of the electroshock devices, a major operational concern is their effectiveness. Electroshock equipment competes with other technologies and use-of-force methods for a space in the arsenal and budgets of security organizations. In the past, the range of electroshock technology has not fared particularly well compared to other options. Austrian researchers investigating the effects of a range of electroshock devices found that the pain inflicted was generally not enough to deter aggressive individuals. The crackling of electricity, though, proved effective as a deterrent.[27] Despite long-standing claims about the effectiveness of TASERs, TASER International has recently stated that its older devices proved ineffective in 15 percent to 33 percent of cases. In response, it has introduced a new advanced line of weapons with significantly higher wattage levels.[28] The new 26-watt version, the ADVANCED TASER, is said to be 99 percent effective in incapacitating individuals, making it more effective than firearms. Promotional material fosters this image by illustrating test results on elite military and police personnel. So, a former US Marines Chief Instructor in hand-to-hand combat states, 'I have been hit by hand grenades yet still completed my mission. The ADVANCED TASER is the only thing that has ever stopped me.'[29]

Of course, determinations of effectiveness depend on the situations in which particularly options are utilized. Contrasting the general effectiveness ratings for electroshock weapons versus batons, for instance, is of limited value. The handful of incidents of TASER deployment noted above by officer Laur consist in the main of a variety of stand-off situations, such as barricaded or suicidal individuals, where many of the contingencies or difficulties surrounding the use of such equipment are minimized.[30] In favor of the effectiveness of such particular equipment, TASER producers have said that their products are often used on those under the influence of drugs like PCP (phencyclidene), a condition that makes individuals much more resistant to control. In addition, there are uncertainties surrounding the extent to which conflict is resolved due to particular pieces of equipment or other factors. Laur mentions a number of cases where the mere threat of using a TASER brought compliance. The latest versions of the TASER, however, have a laser sighting; a device claimed to be an effective tool of compliance in itself.[31]

SOCIAL MOVEMENTS AND CRITIQUE

As in the last section, much of this book has focused on questioning the assumptions and contingencies at work in highly supportive statements of non-lethals. The justification for this has been that these statements are the prominent ones on offer. As already discussed in Chapter 7, electroshock equipment has come under specific criticism by various non-governmental organizations, such as Amnesty International, that are concerned about not only their effects but their (enhanced) potential for abuse. Just what and how such groups substantiate critical claims about non-lethals is worthy of close attention.

Non-lethal weaponry poses various dilemmas for human-rights and civil-liberty organizations. At a basic level, this is a matter of whether or not to spend limited time and resources campaigning about the trade and deployment of such devices, whether to call for reforms of the practices of those committing abuses, or whether to undertake some other approach. The number of individuals working on weapons in general in such organizations tends to be quite small, with even fewer being familiar with non-lethal weapons. Additionally, while there is always the possibility for the abuse of non-lethals, as discussed in Chapter 6, the position of such groups is not simply one of opposition. International agreements encourage the development of non-lethal weapons, and the potential of such technology to minimize injury is widely recognized. Finally, while much of the work of organizations such as Amnesty centers on security forces in countries regarded by many in the West as persistent human-rights abusers and potential torturers, some of the most frequent concerns about electroshock technology relate to the United States, where their deployment is fairly widespread. While calling for an end to the use of electroshock technology in the former is relatively unproblematic, in the case of the latter, the presumption of wide-scale abuse is less easily secured. Likewise, labeling ill-treatment and excessive force in the US as 'torture' is recognized as problematic.[32] In this situation, there are thorny questions as to what sort of analysis should be offered and who will find it convincing.

In 1997, Amnesty International launched its first major study solely dedicated to electroshock technology.[33] It called for the end of the trade of such technology to countries where it was likely to be used for torture or ill-treatment. The report drew attention to those medical studies that noted the potential severe effects and the lack of proper scientific studies. In the group's 2000-1 worldwide torture campaign, Amnesty made other recommendations, including:

- Suspend the use of equipment whose medical effects are not fully known, pending the outcome of a rigorous and independent inquiry into its effects … International transfers should be suspended pending the results of the inquiry.
- Conduct an independent and rigorous review of the use of equipment where its use in practice has revealed a substantial risk of abuse or unwarranted injury. Suspend the transfer of such equipment to other countries pending the results of the review.

- Introduce strict guidelines on the use of police and security equipment [and set] up adequate monitoring mechanisms to keep the guidelines under review and to ensure they are adhered to.[34]

As expressed by such recommendations, Amnesty seeks a legal-rationalistic basis for authority; one that strives for a definitive assessment of the effects and use of technology. The group strives to find general and credible proscriptive measures that in turn make Amnesty International a credible organization. Predictably, correctional officers and others have responded that the controlled use of such weapons (including stun belts) is indispensable in modern, overcrowded prisons and that their merits should be assessed on a case-by-case basis.[35]

The call for the control of the electroshock technology on the basis of definitive assessments and adherence to strict guidelines is understandable but tension-ridden. As already outlined in previous chapters, in general, expert analysis suffers from many limitations: experts disagree, findings are often equivocal, inconvenient analyses are ignored, statements of effects are conditional and relational, and any determinations of safety must be seen in terms of how uncertainties function as part of organizational settings. Specifically in relation to groups such as Amnesty International, there are a variety of problems about drawing on legal-rational evidence.[36] At a basic level, there are the practical problems with appeals to rigorous studies and 'the facts'. At best, resolving the medical effects of a wide range of technologies across varied populations is a time-consuming and expensive process. However, this might not necessarily be a problem for movements if they can muster support for limiting the deployment of a technology until such tests take place. In situations where technology is already in circulation, though (such as much of the United States), the possibility for removing all products until the 'facts of the matter' are established seems remote. That there is generally little support for research into the effects of non-lethal weapons – outside of those state agencies and firms typically promoting the technology – further complicates appeals to 'science'. Perhaps more importantly for human-rights organizations, in seeking a generalizable, definitive basis for assessment constituted by medical studies, there is little space for the subjective experiences of victims (fear, humiliation and pain) as a basis for the recommendations offered.

The call to suspend the trade of electroshock equipment to countries where there is a substantial risk of abuse until the matters are reviewed is also tension-ridden. As explored later in this chapter, there are pressing problems regarding how such information can be obtained, presented and interpreted.

The above commentary is meant to suggest that there are a number of difficulties which non-governmental groups must face in commenting on non-lethals. The author has been a party to many such debates about what needs to be done in Amnesty International (UK): first as an individual member following position developments, then as a (largely observing) member of a committee related to the international arms trade,[37] then moving into 2002 as an active contributor to policy debates about non-lethal weapons. In these roles, I have struggled with others regarding the tensions mentioned above. As with other organizations, groups such as Amnesty try to advance generalized claims that provide a credible basis for policy positions while noting the variations in the effects of technology depending on situational particulars. In this there are tensions of whether to make general calls that might capture the media and public's imagination versus acknowledging the relational and situational merits of technology. Just what statements might count as credible in the eyes of governments and others is an issue that looms large. The language of officialdom is perceived as one of objective, scientific and definitive facts – even if such demands are unrealistic.

In perceiving the situation as such, Amnesty as an organization only takes a position on weapon transfers when they can be tied to documented human-rights violations. While it might on occasion note with some disapproval the possible example established by the uptake of electroshock weapons (as in Chapter 7: see p. 153), such remarks do not shape core policy recommendations. As such, various issues associated with non-lethals receive little comment. While some members of the organization might have concerns extending far beyond the safety of specific devices or particular instances of abuse (say with regard to non-lethal weapons research possibly leading to the 'industrialization' of human-rights abuses), such issues do not manifest themselves in policy recommendations. In studying non-governmental organizations and public-policy controversies, Nelkin has argued that scientific and technical concerns about risk often function as a surrogate for more widespread social and political issues which motivates controversy.[38] The experience of the author would certainly affirm this argument. Calls for a clarification of the facts of the matter and appeals for more evidence are certainly relatively easy to make and appreciated by the media, even when those making them recognize the issues at stake are far more complicated.[39]

MARICOPA COUNTY JAILS

This section elaborates the themes raised so far through an examination of a major non-lethals initiative in the United States: the introduction of

stun guns and pepper sprays for all detention officers in Maricopa
County Jails. Maricopa County Sheriff's Office (MCSO) operates the
jail system that serves Phoenix, Arizona and the surrounding area. The
jail system includes short-term (up to a year) holds and a central 'Intake'
facility for individuals on very short-term sentences and those awaiting
trial.

Although some non-lethal weapons (such as chemical irritants) have
been widely available in US prisons and jails for some time, at most they
were distributed to senior command officers. For advocates of the technology, this situation was quite limiting. Starting in 1994, the National
Institutes of Justice and the National Sheriffs' Association funded the
provision of non-lethal stun guns and pepper sprays to all detention
officers. The sheriff of the County, Joe Arpaio, was highly optimistic of
the potential of this pilot study:

> The Maricopa County Sheriff's Office is all but abandoning
> physical force as the primary way to restrain unruly prison-
> ers and instead will rely on non-lethal pepper sprays and stun
> devices that promise reduced injuries to both lawmen and
> criminals ... when suspects or jail inmates refuse to respond
> peacefully to lawful instructions, the pepper spray or stun
> device certainly is more efficient and humane than heavy
> physical force.[40]

Non-lethals were to be used early and fairly frequently in conflict situations, even where inmates only passively resisted officers' commands.
As an 'alpha test site' for the potential widespread use of non-lethals,
the Office recognized that the initiative would 'undoubtedly attract
national attention, and presumably influence decision making in many
Sheriffs' Departments'.[41]

Arizona State University

As part of the pilot, criminologists at Arizona State University
(Hepburn, Griffin and Petrocelli) undertook an official in-depth evaluation study covering the time from the introduction in 1994 to 1996.[42]
As mentioned in previous chapters in relation to OC sprays, such evaluations have played a major role in shoring up particular evaluations of
the merits of non-lethals. Published in September 1997, the report drew
on a number of key sources: altercation forms specifically introduced
for the pilot that were supposed to be filled in after every use-of-force
incident, three longitudinal surveys of all detention officers and selected
interviews. From the start, the remit of the analysis was limited. As the

authors state, it 'does not attempt to address questions pertaining to whether the use of force in any particular situation is appropriate or inappropriate' and not whether it is within the official guidelines. 'Instead we focus our analysis on whether the use of non-lethal weapons achieved the desired outcomes of more effectively controlling inmates while also reducing injuries to officers and inmates.'[43]

On the basis of the data collected, the authors constructed a diverse array of graphs, charts and correlations meant to provide conclusive indicators of the merits of non-lethals. The overall story told is one of officers moving from having 'serious reservations' about non-lethals to being gradually convinced of their utility through strong leadership and comprehensive training. Other main findings included:

- Nonlethal weapons have become an integral tool in the officers' response to altercations with inmates. Nonlethal weapons appear to be appropriate for nearly two-thirds of all altercations, and they were used in nearly half of all altercations. Further support for the importance of the weapons is found in the fact that nearly half of all uses of the weapon require only a display or threat and not an actual application to the inmate.
- The stun device was quickly adopted and frequently used in altercations. Appropriate for most situations, the stun gun was used in more than half of all altercations ...[44]

And yet, as with so many evaluation studies, the ability to make unequivocal assessments was acknowledged as limited. The findings were somewhat inconclusive or conflicting on main points and the movement from the data to conclusions required various bridging assumptions. So, in contrast to claims about non-lethals reducing injury, the percentage of inmates that recorded injuries in use-of-force techniques was lowest from pepper spray (8 percent), then ordinary hands-on tactics (~ 13 percent), and then stun guns (~19 percent). However, overall, officers saw pepper spray as the least effective option whereas hands-on tactics were the most effective. The rate of injuries to officers at Intake was lowest from hands-on tactics, figures that the authors conjectured were 'probably a function of the increased number of officers available to control the inmate'. In general, little data was gathered on the types of injuries, their severity or the duration of their effects. There is no indication that officers or inmates were monitored in the medium- or long-term for injuries (say in relation to multiple exposures from the sprays or stun guns). Past examples given in this book would suggest that while this lack of monitoring is hardly unprecedented, it does little to encourage confidence in the robustness of findings. The report was also unable to determine that assaults on officers were reduced, despite MCSO claims that altercations reduced since the start of trials.[45]

Depending on one's starting position, hype rather than experience appears to have been a key part of the assessments made. So, Hepburn, Griffin and Petrocelli[46] found in the central Intake facility that '[half] of all officers responding to the third survey in 1996 stated that the pepper spray and the stun devices were frequently or always needed ... This is a rather strong endorsement of non-lethal weapons, especially in light of the fact that the pepper spray or foam was virtually never actually used within Intake.'

Within the report's text, there are reasons for doubting the validity of findings and some of the statements given above. For instance, at the start of the pilot, the authors stated that 83.1 percent of officers believed non-lethals would reduce injuries to officers. That percentage remained at roughly that level throughout the pilot. While this number is high, it is somewhat at odds with the injury data mentioned above in relation to stun guns. As with the case of CS sprays in the last chapter, there are discrepancies between injury records and officer-surveyed views that can be read in any number of ways. In addition, the percentage of officers who agreed that stun devices are frequently or always needed, as well as officers' belief that the availability of weapons would affect inmate misconduct, slightly declined over the course of the pilot.

With a central focus regarding the effectiveness determination of users, the report makes little room for the views of recipients. The extent of questionable applications was determined through grievance forms against officers filed by inmates (though these forms were not available at Intake). Despite claims not to comment on the appropriateness of particular uses of force, the authors took the lack of rise in grievance claims by inmates during the pilot as evidence that 'the officers did not use, or abuse, the weapons excessively'.[47] Further along these lines, it was concluded that '[t]here were few reported instances in which a nonlethal weapon was misused or was used to abuse an inmate, and disciplinary action was taken against the officer in each known case'.[48] The report mentions that the US DoJ started an investigation into allegations of excessive force in Maricopa County jails in August 1995. This resulted in a letter to the Jails' Board of Supervisors in March 1996. Hepburn, Griffin and Petrocelli noted a substantial decrease in altercation reports after the federal investigation began in 1995. Regarding this, they state: '[t]he sudden decrease in reported altercations in August 1995 suggests that these events affected either: (1) the number of incidents in which force was used or threatened or (2) the likelihood that an altercation form would be completed follow-

214

ing a use-of-force incident, or (3) both'.[49] No further implications are drawn out regarding the validity of altercation reports.

Department of Justice – Evaluation I

Penologist Eugene Miller conducted the first DoJ investigation, which was published in a report in January 1996.[50] The full report was obtained by the author through the US Freedom of Information Act. It drew on data collected from inspection tours, formal and informal interviews with prison staff, interviews with pre-selected and random inmates, 'Use-of-Force' reports, as well as various internal documentation and video footage. In contrast to the account of the jails given above, Miller was highly critical of the use of non-lethals and found there was 'definitely evidence of on-going use of excessive force with Maricopa County Jail system'.[51] He made a series of statements about the way allegations were handled: the procedures were not independent or robust, complainants were not often interviewed, and the frequency of allegations was not monitored. The altercation form, believed to have been introduced as part of the Arizona State University study, was said to be a mere checklist and 'particularly uninformative'.

Miller located the problems associated with force in jail conditions. Many detention officers were young and inexperienced and there was a general understaffing due to the recent boom in the jail population. In this situation, he argued, an inadequate number of youthful and inexperienced officers felt unduly threatened. Instead of merely extending the capabilities of individual officers, the stun guns and pepper sprays served more as substitutions for missing staff. This was not seen as the fault of the training staff, but rather reflected the profile of detention officers and the unruly and overcrowded jail system. The easy availability of non-lethals in this context transformed officers' responses to situations. So, the non-lethals became compliance tools for maintaining order rather than use-of-force responses to violent acts.

As with so much surrounding non-lethals and the use of force, the 'facts of the matter' were highly disputed. The initial report is written in a claims–counterclaims format that juxtaposed Miller's claims with responses from MCSO. For instance, he contends that there were

> ... numerous instances of questionable or apparently inappropriate use of non-lethal weapons. An incident that illustrates the danger of all detention officers to have non-lethal weapons was brought to my attention by a senior staff member: An 18 year-old detention officer on his last day on the job decided to use his stun gun on an inmate just to see how it would work.

> RESPONSE [from MCSO]: The writer states there were 'numerous instances of questionable or inappropriate use of non-lethal weapons' but does not provide specifics so these can be addressed. If the investigative team had brought these allegations to the attention of the appropriate Division Commander at the time they witnessed them, it would have been far easier to provide a reply.[52]

MCSO went on to add that it had no knowledge of the specifics of the case mentioned and that, in fact, during the time period under question, no 18-year-old officer finished his last day on the job. The Office refuted Miller's allegations, claiming they were based on inaccurate or incomplete information.

Just as the diagnosis offered was a source of disagreement, so too was the remedy. Miller contended that the inappropriate use of non-lethals was 'inevitable' when deployed so widely. He recommended that only floor supervisors should have these devices. MCSO cited a new tough disciplinary program to combat excessive use of force. The Office insisted that 'the program has worked exceptionally well. The officers realize the non-lethal weapons must be used judiciously within the detention setting. We have found the use of non-lethal weapons, as an alternative to hands on force, is reducing the number of injuries to both staff and inmates.'[53]

Amnesty International

After receiving numerous allegations of excessive force, in June 1997 Amnesty International visited Maricopa County jails. In August of that year, it launched a report about ill-treatment covering a number of aspects of the jails including use of stun weapons and pepper spray. Amnesty drew on the 1996 summary of findings of the DoJ which concluded that the ease of availability of weapons was a substantial problem that led to unconstitutional conditions.[54] In contrast to the claims made above by the Arizona State researchers, the group pointed to the failure to discipline officers for improper behavior. Again in contrast, various prominent allegations about the excessive use of force against inmates were listed:

> One example is the case of Richard Post, a paraplegic who was admitted to the jail in a wheelchair in March 1996 and alleges that he was placed in an isolation cell for an hour without medical attention, despite asking for a catheter so that he could empty his bladder. He tried to seek attention by

banging on the cell window and eventually blocked the toilet in the cell, causing water to seep under the door. Detention officers then removed him from his wheelchair and strapped him into a four-point restraint chair, with his arms pulled down towards his ankles and padlocked, and his legs secured in metal shackles. He claims that straps attached to the chair behind his shoulders were tightened round his chest and neck so that his shoulders were strained backwards, and that one guard placed his foot on the chair and deliberately yanked on the strap as hard as he could. It is further alleged that an officer threatened him with a stun gun while he was immobilized in the chair, while other officers looked on. For the first hour that he was in the chair he was denied the gel cushion he had with him, with the result that severe decubitus ulcers developed around his anus. The manner of his restraint is reported to have caused compression of his spine and nerve damage to his spinal cord and neck, resulting in significant loss of upper body mobility.

Although no use of force report appears to have been made in this case, an internal inquiry was held after his mother complained about his treatment. The sergeant who took the decision to place Richard Post in the restraint chair said that this was done for his own safety as he had been banging on the cell window threatening to harm himself.[55]

Consistent with the search for definitive assessments of particular situations discussed in the last section, Amnesty recommended MCSO conduct stringent reviews of allegations of misuse of stun guns and reassess the use-of-force policy. What it would take for such actions to be convincing was not specified. Almost no reference was made in the report to possible limitations stemming from how this information was gathered.

Before proceeding to consider the second DoJ investigation, it is worth considering how human-rights violations are reported. For Amnesty and others without open access to jails, there were significant questions about what claims should be made, how these are sustained and what implications should follow. As a basis for claims to legitimacy, much of the weight of human-rights reporting comes from speaking for those whose voices are usually not heard. Doing so is not unproblematic. Just how this ought to be done has been a source of much discussion among activists, academics and others. In this regard, anthropologist Richard Wilson has commented on the contingencies and limitations of reports of human-rights abuses.[56] He argues that the legal-rationalistic basis for authority typically sought by human-rights groups aspires to a 'culture of scientism' as represented by the search for universal classification and objective data. So, as in the case above, human-rights organizations typically adopt a legalistic language to describe individuals wronged in particular situations. Specific allegations of abuse consist of a litany of actions whose motives appear incomprehensible.

Wilson contends that, in practice, the extent and type of reporting of violations is contingent on a number of issues related to the perceived nature of offences, the targets, and the sources of information at the disposal of organizations. Herein, '[i]nstead of a documentary style which recognizes the indeterminacy of a case (which human rights organizations generally recognize at a different level) and the limitations of any media representation, the facts in the main text of human right reports simply speak for themselves'.[57] Whatever, difficulties were experienced and acknowledged in gathering information are not incorporated into recommendations or prominently acknowledged. Rather, the attention is given to human-rights violations that are in need of immediate action. While human-rights organizations might often draw on subjective accounts by victims of pain, humiliation, fear and uncertainty, in the striving for scientism and 'actionable certainties', subjectivity is restricted to victims rather than to the organizations decrying abusive actions.

Wilson argues that there is significant debate among human-rights activists and others about just what sort of accounts should be given. In simplified terms, the debate is characterized as between legalists and contextualists. At stake is whether reports only provide the facts of particular cases, or try to situate allegations in a wider social, economic and political context so as to be able to make greater sense of actions.

What might such a context be in this case? MCSO has gained national and international notoriety in recent years because of its hard-line stance on the treatment of prisoners and inmates. Sheriff Joe Arpaio goes by the title of 'America's Toughest Sheriff'. This image of toughness is actively encouraged. As promotional material for the MCSO states, '[n]o other detention facility in the country, state or county, can boast of 1200 convicts in tents; no other county or state facility can boast of a gleaning program that results in costs of under 45 cents per meal per inmate; few others can say they have women in tents or on chain gangs …'.[58] Other activities include the provision of old-style black and white inmate strips and issuing of pink underwear.[59] With the spotlight that has followed such measures, numerous allegations of corruption, the misuse of force, and other wrongdoings by the MCSO have been made.[60] In adopting a legalistic line, Amnesty did not draw attention to this 'get-tough' culture or other related concerns.

Department of Justice – Evaluation II

On the basis of counter-commentary from the MCSO about the Miller investigation, during March/April 1997 another DoJ examiner, George Sullivan, was appointed to conduct a second review.[61] Like Miller, Sullivan based his claims on various MCSO documentary evidence as well as staff and prisoner interviews. Like Miller, Sullivan found concerns about the unnecessary and excessive use of force. Finally, like Miller, he cautioned against endorsing non-lethals for all detention officers. Furthermore, Sullivan obtained a security company's research report conducted by Mike Doubet (see the fourth section of Chapter 5, p. 105), and concluded that all pepper sprays should be removed immediately and their use forbidden until considerably more attention had been given to their necessity and desirability.

As opposed to previous evaluations, Sullivan's major concerns rested with the liberal use-of-force policy in the jails. Initially, given the promotion of the non-lethal as safer than hands-on techniques, it was justified against inmates who resort to passive resistance. With such an endorsement, he maintains, non-lethals were used as compliance tools. Eventually, after various allegations of excessive force, pepper sprays and stun guns moved from being options for use against passive resistance by inmates to being ones not justified for passive or active resistance. While these developments were noted in the Arizona State evaluation, in no way did that study attempt to relate changes in the use-of-force policy to injuries or grievances filed. Yet, for Sullivan, this was the key issue.

Also cited as important contributory factors in 'what went wrong' were the continuing inadequacies of the inmate-complaints procedure and the general 'get-tough' approach within Maricopa jails. With regard to the former, the difficulty of taking forward grievances produced a 'chill factor' in respect to making allegations. The latter referred to the general 'get-tough' culture in the jail system. Unlike with Amnesty International and others, the context of the 'get-tough' culture was taken as 'regrettable', and as an important contributor to the excessive use of force with non-lethals. Sullivan stressed the need to differentiate between the public rhetoric given about the desirability of harsh incarceration conditions versus the sort of advice that should be given to officers about the appropriateness of force.

Much fanfare surrounded the pilot study in Maricopa County jails. Arguably, the initiative has not worked out as originally planned. One month after the release of the Arizona State University report and two months after the Amnesty report, in October 1997 the DoJ came to an agreement with MCSO regarding how to reduce excessive force in the jails. The measures included:

- Implementing the Jail's new policy restricting the use of non-lethal weapons such as stun guns and pepper spray;
- Continuing implementation of a Use-of-Force committee to review all allegations of the use of excessive force and restraints;
- Changing the Jail's grievance system to make it easier to file excessive force and restraints complaints, and changing the way in which resulting investigations are conducted.

As elaborated in the last section, behind such proclamations lay multiple ways of making sense of what happened. Had it not been for DoJ and Amnesty International investigations examined in the previous section of this chapter, the Arizona State University evaluation would have stood as the authoritative account of what happened during the trial. Despite the contentions surrounding what happened and why, this has not stopped NIJ-sponsored investigators from quoting bald figures about optimistic findings of officer satisfaction in the Arizona State evaluation as evidence for the merits of non-lethals.[62] Against this backdrop, the Sheriff of MCSO has exonerated officers and claimed he has not changed use-of-force practices since the DoJ settlement.[63]

The remainder of this chapter draws out a number of lessons from the foregoing analysis of the Maricopa County trial. First, it seems unrealistic to suppose that the debate about what really happened at the jails could be resolved simply by the provision of more information, the undertaking of a few additional interviews or the summing-up of opposing viewpoints. Accounts of excessive force are disputed on the substance of evidence, the source of evidence, the interpretation of incidents, the accuracy of descriptions and the up-to-dateness of information. The alternative appraisals previously mentioned indicate the importance of the interpretation of evidence. Hepburn and colleagues approvingly found that merely threatening inmates with the use of stun guns served as an effective means of control in 27 percent of situations. While this aspect of the introduction of the technology might be reassuring for detention staff, in the context of allegations of human-rights abuse, such findings have wholly different implications for others. Amnesty International, and even the DoJ investigators, would no doubt find reason for concern when 92.5 percent of officers reported prisoners were more afraid of stun guns than hands-on control techniques, and 63.7 percent said the devices gave them more authority. The pursuit for quantified figures offers little hope for resolving such disputes. What sustains the argument is more than the clarification of the facts of the matter or the lack of robust indicators.

Perhaps, more fundamentally, the process for defining, identifying and interpreting the issues at stake and determinations of what constitutes acceptable evidence inform alternative accounts. MCSO sought to ensure that credible assessments were based on the sort of detailed and insider knowledge which it held about jail procedures and conditions, when access to such information was clearly limited. Amnesty adopted a legalistic approach to justify credibility of its claims. Other commentators have taken a different orientation to what constituted credible claims, given disputes about the jails. Investigations by a local newspaper, the *Phoenix New Times*, took the disputes and limitations of obtaining details about jail practice as central topics of analysis and critique. In the case of Richard Post, for instance, Ortega outlines alternative accounts between Post and the detention officers; the alleged selective, doctored and delayed release of relevant jail closed-circuit videotapes; and the inability of video images alone to resolve questions about verbal abuses and threats.[64] Allegations of cover-up, the withholding of evidence and the destruction of evidence have been made in relation to other cases.[65] Taking such reports seriously means that calls for, or claims to speak about, the facts of the matter of the trials are highly problematic.

So, what really happened during the trials? How widespread were concerns about the excessive use of force with non-lethals? For pragmatic purposes, weighing up the evaluations given, it might be concluded that there had indeed been basis for concern about excessive force with non-lethals. Certainly, in the end, the responsible authority (DoJ) came to that conclusion. But even if we accept that something went wrong, the question remains, why? What lessons do we learn? What specifically might this case tell us about the appropriateness of non-lethal weapons?

Contexts and Technologies

In the remainder of this chapter, I wish to reiterate and extend some of the themes raised already in this book. As mentioned above, there are debates in human-rights quarters and elsewhere about how to report allegations of human-rights violations, whether by providing minimalist descriptions of events or by situating allegations in relation to a wider context. 'Contextualists' call for a greater incorporation in reports of the social, economic and political context in which human-rights violations take place. The overall suggestion is to provide a narrative that gives some sense of meaning or explanation to events.

Of course, in many respects, just how accounts should be offered depends on the audience one is speaking to and the purposes sought. As argued above, though, legalistic and highly decontextualized accounts offer little assistance in helping understand the range of issues at stake in considerations about legitimacy and non-lethal weapons. However, attempts to provide the appropriate contexts for understanding what happened during the trial is likewise fraught with problems. So, in just what context ought we to understand the trial? Is that the context of chronic understaffing in the jails due to rising prison populations, the context of the 'get-tough' culture, or something else? The alternative evaluations given above illustrate how different contexts are drawn on to give distinctive interpretations of what happened. These, then, have significant implications for the credibility of evidence cited. In contrast to the Arizona State evaluators, the DoJ investigators did not have much faith in inmate grievance forms filed as any real indicator of the scope of excessive force by officers because of the defects identified about the operation of the complaints system.

But, as before, this is not merely to suggest that technology must be seen in relation to context, even if we accept that there might be multiple and competing conceptions of what that might be. Rather, weapons take their meaning from an understanding of their context and, at the same time, the context develops from an understanding of the technologies. So the Arizona State University investigators struggled with how to account for the findings that the level of injuries with stun guns was higher than that for typical hands-on tactics. They did so by drawing attention to presumed situational factors surrounding the use of these electroshock devices. Likewise, the reason for which the lowest rate of injury to officers took place with hands-on tactics at Intake was seen as 'probably a function of the increased number of officers available to control inmates'. So, where non-lethals are recorded as incurring greater injury than previous methods of control, or when officers experience lower rates of injury from previous methods (both at odds with the stated rationale for the weapons), this is due to situational factors rather than from the technology 'itself'. The context is defined in relation to the technology because an understanding of the weapons tells you what contextual factors are worth considering. For instance, past arguments made in this book give reason for suggesting that pepper sprays and stun guns might have greater effects on people taking drugs or on those sprayed or shocked repeatedly. Such assertions and their denial lead to alternative proposals about what specific surveillance procedures should be in place and what sort of injuries should be recorded. In other words, what we take as the technology tells us which features or lack of features in the context we are looking for in evaluations. In this regard, evaluations are not simply a matter of specifying their effect and then focusing on their use, but thinking of the two reciprocally defined.

Another difficulty of drawing on context in formulating a narrative of events is that such stories often rely on assumptions about the intentions and motivations of actors to inject a sense of meaning to events. A key question in evaluating these complex situations is whether *post hoc* or general ascriptions of intent are helpful. As I have suggested in a number of chapters, the importance of this mutual definition of technology and context is that there is a constant displacement in disputes about who or what is to blame, whether that be users, technology, situations of use, etc. The importance given to these factors is defined in their relation to one another depending on analysts' 'sociology' of disputes. So for Miller, many of the problems with the use of force at the jails stemmed from the 'flawed' character of front-line guards (being youthful and inexperienced) who were working in demanding environments. In these conditions, the introduction of non-lethals for all officers encouraged an excessive use of force. In the language of discussions in Chapter 4, the non-lethals played a *transformative* role. For Sullivan, the introduction was more *additive* in its implications, because the major source of trouble was the lax rules provided by senior officers. In the Arizona State University study, the views of officers are taken as the basis for the evaluation made and are therefore implicitly beyond reproach. In Amnesty International's account, the actions of officers were largely incomprehensible. As with claims about the relevant context for evaluation, attributions of intent or the focus on intent at all (whether that be seen as 'good' or 'bad' – see Chapter 8 on pp. 193–8), are interpretative devices used to render events understandable. Depending on how motivations are attributed, alternative recommendations are offered about what (if anything) needs to be done to remedy situations, whether that be alterations to the weaponry used, to training provisions, to physical workplace conditions, or to something else altogether.

Of course, all of these points above about the selective drawing on context and the marshaling of evidence to imply a sense of motivation apply to the 'meta' account given in this chapter of the various evaluations. In no way has it presented anything like a full and uncontestable mere description of what really happened during the trials, or given the proper context in which to understand the various evaluations. This chapter is itself a mobilization of particular contexts to fashion a story together to suggest an interpretation. It is open to question, as such, regarding its coherence, applicability and plausibility of the argument offered.[66] Throughout, choices were made as to which points to draw on and how to present them. Those able to get a hold of the central documents of this chapter are encouraged to formulate their own account of what happened.

Importance of the Knowledge at Hand

Continuing with a consideration of lessons from Maricopa County, determinations about how the facts ought to be read in evaluation studies depend on the knowledge held as well as knowledge of how knowledge is distributed. For instance, the extent of injuries recorded from the use of force in the jails depends on the facilities in place to monitor it. This dependence is perhaps particularly acute for the pepper sprays and stun guns as opposed to hands-on tactics because they are more likely to lead to injuries with which jail medical professionals would have had little previous experience, and because there might be disputes about whether medium- and long-term ailments derived from these devices.

Marshaling yet another rendition of the context, it is possible to further underline the importance of the knowledge at hand in relation to injuries. In 1999, as a result of a four-year investigation by the Justice Department, Maricopa County reached a second agreement with it regarding the need to reform jail practices. The DoJ found the jail system had failed to provide adequate medical care for inmates.[67] Among the areas seen as in need of improvement was that of how inmates' requests for medical treatment and physical examination were handled.[68] Medical screening for health conditions at Intake – the sort of procedure necessary for some mentioned in Chapter 5 to mitigate the dangers associated with OC sprays – was seen as particularly wanting. The findings of this separate investigation cast some doubt about the validity of the injury rates and assessments of injury recorded for non-lethal weapons. This second investigation was not concluded until after all the evaluations discussed above had ended. None of the evaluation studies gave attention to the adequacy of health provisions in relation to determination of the rate of injuries and the acceptability of non-lethals.[69]

Non-lethals are supposed to be a means of minimizing controversy surrounding the use of force through the application of minimum harm. But, as suggested, attempts to legitimatize intervention through such weapons are open for dispute. Just as the basis of claims about the safety or danger of the technologies in Chapter 5 could be unpacked to indicate something of the contingencies and uncertainties underlying particular evaluations, so too it was argued that this can be done for electroshock technology. In the case of the trial of pepper sprays and stun guns in Maricopa County, the conflicting accounts of what went on during this exercise provide a means of bringing to the foreground pertinent questions about the credibility of claims, the interpretation of evidence, and other issues at stake in how and what evaluations are offered.

NOTES

1. For a further discussion of non-lethals in prisons, see Mampaey, L., *Prison Technologies*, Report for the Scientific and Technological Options Assessment PE 2889.666 (Luxembourg: European Parliament, 2000).
2. Heal, C., 'The Evolution from "Non-Lethal" to "Less-Lethal"' [cited 20 March 2001]. See www.airtaser.com/Web_2000/Feb/SidHeal.htm
3. Smith, R., 'Reducing Violence', in Alexander, J., Spencer, D., Schmirt, S. and Steele, B. (eds) *Proceedings of Security Systems and Nonlethal Technologies for Law Enforcement*, (Boston: The International Society for Optical Engineering, 1997), 27.
4. Smith, R., Personal Communication with author, 25 January 2000.
5. Ordog, G., Wasserberger, J., Schlater, T. and Balasubramanium, S., 'Electronic Gun (Taser) Injuries', *Annals of Emergency Medicine*, 16 (1987), 73–8.
6. Kornblum, R. and Reddy, S., 'Effects of the Taser in Fatalities Involving Police Confrontation', *Journal of Forensic Sciences*, 36 (1991), 434–8.
7. TASER International. See www.airtaser.com/Med%20Studies/overview.html
8. J&L Self Defense Products [cited 23 Jan. 2001]. See www.selfdenfenseproducts.com
9. Douez, S., 'Police Force Looks at New Weapon, and it's a Stunner', *The Age*, 10 January 2001.
10. Laur, D., *Independent Evaluation Report of TASER and AIR TASER Conduced Energy Weapons* (Victoria: Victoria Police Department, 1999).
11. See Hall, S., 'Culture, Media, and the Ideological Effects', in J. Curran et al. (eds) *Mass Communication and Society* (Beverly Hills: Sage, 1977), 315–48.
12. Roy, O. and Podgorski, A., 'Tests on a Shocking Device', *Medical & Biological Engineering & Computing*, 27 (1989), 445–8.
13. Ordog et al., 'Electronic Gun (Taser) Injuries'.
14. Kornblum and Reddy, 'Effects of the Taser in Fatalities Involving Police Confrontation'.
15. Allen, T., 'Effects of the Taser in Fatalities Involving Police Confrontation', *Journal of Forensic Sciences*, 37 (1992), 956–8.
16. Ibid., 957.
17. Orlando Sentinel Tribune, 'Stun Gun Target Awarded $225,000 for Miscarriage', *Orlando Sentinel Tribune*, 16 June 1991, A6; Cusac, A., 'Stunning Technology', *The Progressive*, July 1996; Doucet, I. and Lloyd, R., *Alternative Anti-personnel Mines* (London: Landmine Action and German Initiative to Ban Landmines, 2001); Loviglio, J., 'US: Want Stun Guns Safety Issue Raised', *Associated Press Report*, 20 February 2002 and Hammack, L., 'Conn. Inmate at Va. Prison Died after Shock', *The Roanoke Times*, 17 May 2001.
18. O'Brien, D., 'Electric Weaponry', *Annals of Emergency Medicine*, 20, 5 (1991), 583–7.
19. Omega Foundation, *Crowd Control Technologies*, Report to the Scientific and Technological Options Assessment of the European Parliament, PE168.394 (Luxembourg: European Parliament, 2000).
20. Jafari, H., Couratier, P. and Camu, W., 'Motor Neuron Disease after Electric Injury', *Journal Neurol Neurosurg Psychiatry*, 71 (2001), 265–7.
21. My thanks to correspondence with W. Camu for clarifying this possibility.

22. Bleetman, A. and Steyn, R. *The Advanced Taser*, 17 December 2000, 1–27.
23. Bleetman, A., Personal Interview, 7 December 2000.
24. Ibid.
25. Bleetman and Steyn, *The Advanced Taser*, 20.
26. See also Brooks, M., 'Shock Tactics', *New Scientist*, 11 August 2001, 11.
27. Denk, W., Missliwetz, J., Tauschitz, C. and Wieser, I., 'Electric Shock Devices as Weapons', in *Non-Lethal Weapons*, Proceedings of the 1st European Symposium on Non-lethal Weapons, 25–26 September 2001, Ettlingen, Germany (Posfach: ICT, 2001).
28. TASER International. See www.airtaser.com/Med%20Studies/overview.html
29. TASER International, *This is as Close to 100% TAKEDOWN POWER as You Can Get*, promotional literature (Scottsdale, AZ: TASER International, 2000).
30. Laur, D., *Independent Evaluation Report of TASER and AIR TASER*.
31. Houde-Walter, W., 'Violence Reduction and Assailant Control with Integral Laser Sighted Pistols', in *Proceeding of Security Systems and Nonlethal Technologies for Law Enforcement* (Boston: The International Society for Optical Engineering, 1997).
32. Mecklin, J., 'Barbarism as a Public Relations Strategy', *Phoenix New Times*, 5 December 1996. See www.phoenixnewtimes.com/
33. Amnesty International, *Arming the Torturers*.
34. Amnesty International, *Stopping the Torture Trade* (London: Amnesty International, International Secretariat, 2001), 51.
35. Hinman, L., 'Stunning Morality', *Criminal Justice Ethics*, Winter/Spring (1998), 3–13.
36. For a broader analysis see Yearly, S., 'Green Ambivalence about Science', *British Journal of Sociology*, 43, 4 (1992), 511–32.
37. To provide some context, starting in 1997, the author was a member of the Military, Security and Police Working Group. The Group's primary aim is to track and campaign for the end of the proliferation of military and police weapons (whether 'non-lethal' or 'lethal' but generally the latter) from Britain to countries where they are likely to be used in human-rights violations.
38. Nelkin, D., 'Science Controversies', in S. Jasanoff et al. (eds), *Handbook of Science and Technology Studies* (London: Sage, 1995).
39. For an instance of the author falling into this see: Hopkins, N., 'Stun Guns on UK Torture List', *The Guardian*, 2 August 2001.
40. Arpaio, J., 'Non-lethal Weapons: The Beginning', *Roundup*, February, in Hepburn, J., Griffin, M. and Petrocelli, M., *Safety and Control in a County Jail* (Tempe, AZ: Arizona State University, 1997).
41. Sullivan, G., *Report of Corrections Consultant on the Use of Force in Maricopa County Jails Phoenix, Arizona*, May 14 (Washington, DC: Department of Justice, 1997), 15.
42. Hepburn et al., *Safety and Control in a County Jail*.
43. Ibid., 8.
44. Ibid., 52.
45. Amnesty International, *Arming the Torturers*.
46. Heburn et al., *Safety and Control in a County Jail*, 31.
47. Ibid., 45.
48. Ibid., 17.
49. Ibid., 6.
50. Miller, E., *Response and Outline to Expert Penologist's Report: Use of Force in Maricopa County Jails System* (Washington, DC: Department of Justice, 1996).
51. Ibid., 30.
52. Ibid., 7–8.
53. Ibid., 6.
54. Department of Justice [DoJ], *Civil Rights Division Letter to Maricopa Board of Supervisors*, 25 March (Washington DC: Department of Justice, 1996).
55. Amnesty International, *Ill-treatment of Inmates in Maricopa County Jails, Arizona*, AMR 51/51/97 (London: Amnesty International, 1997).
56. Wilson, R., 'Representing Human Rights Violations', in R. Wilson (ed.), *Human Rights, Culture, and Context* (London: Pluto, 1997).
57. Ibid., 143.

58. Maricopa County Sheriff's Office (MCSO), 'About MCSO' [cited 14 March 2001]. See www.mcso.org/submenu.asp?file=aboutmcso&page=main

59. The latter was justified to prevent discharged inmates from taking white underwear when they left the jails. The commercial selling of this underwear to the local population has apparently become a profit-making venture. Those interested in reading about this and others legends of MCSO are advised to visit www.posse.net or www.crime.com

60. See the archive of *Phoenix New Times* at URL http://www.phoenixnewtimes.com/ for such allegations.

61. Sullivan, *Report of Corrections Consultant*.

62. Smith, M. and Alpert, G., 'Pepper Spray', *Policing*, 23, 2 (2000).

63. Ortega, T,. 'Fed Up', *Phoenix New Times*, 15 April 1999. See www.phoenixnew times.com/

64. Ortega, T., 'Lies and Videotape', *Phoenix New Times*, 15 April 1999. See www.phoenix newtimes.com/

65. Maximum Films, *America's Toughest Sheriff* (London: Maximum Films, 2001).

66. For a much wider discussion of the points raised here see Herrnstein Smith, B., *Belief and Resistance* (Cambridge, MA: Harvard University Press, 1997).

67. See Anon, *Issues Pertaining to Medical Services at the Maricopa County Jail Correctional Health Services: Prepared for the United States Department of Justice Civil Rights Division, Section on Special Litigation*, 7 February 1997.

68. Department of Justice, *Maricopa County to Improve Medical and Mental Health Care for Inmates*, 99-588, 6 December (1999).

69. It should be noted, though, that the Arizona State University evaluation was the only one that specified injury-rate figures.

10

Humanitarian Interventions, Humanitarian Tools?

Much of the detailed discussion of non-lethals so far has dealt with 'first-generation' devices, at the 'low-tech' end of the spectrum. In this, the kinetic munitions, electroshock weapons and chemical irritants examined have functioned as tactical aids in dealing with particular situations. This chapter shifts the focus by turning toward speculative future deployments of non-lethals and their stated potential for bringing about major alterations in the way conflict is handled. As elaborated earlier, much of the current promise relates to the possibility of getting beyond the 'rubber-bullet' type option.[1] Especially for military operations such as humanitarian interventions, the hope for some is that the availability of new weapons will help usher in different ways of resolving conflict.

This chapter examines various prominent scenarios, including those by the US Marines and the US Council on Foreign Relations. In keeping with the desire to ask what claims attract the most support and why, this chapter speculates about the likely terms for debates about the merits of future non-lethals in humanitarian interventions. Following earlier arguments, I maintain that we should understand the deployment of this weaponry as bound up with attempts at control; controlling individuals' behavior but also controlling assessments of events. Securing the sort of promise attributed to non-lethals will depend on strategies for managing interpretations, rather than merely as the result of particular choices about how technologies are employed.

PEACEKEEPING AND HUMANITARIAN INTERVENTIONS

It is often argued that since the end of the Cold War, the bipolar divisions that imposed order in the past have been eroded. With this has come a greater enthusiasm and willingness for international interventions within (at least) certain state boundaries for the purposes of various peace support operations, such as humanitarian assistance.[2] The military operations in Northern Iraq after the Gulf War, the Balkans, Rwanda, Haiti, Somalia and Sierra Leone are some prominent examples of such missions. Of course, military campaigns throughout history have been justified by appeals to 'humanitarianism'. One major reason cited for the need for intervention today is the increasing rate at which civilians are killed in modern conflicts.[3] In societies divided along factional and ethnic lines, the increasing use of non-combatants as pawns and targets in power plays between rival groups has drawn much condemnation. Multilateral actions by organizations such as the UN or NATO are typically justified not in terms of the direct national interests of particular countries, but by the need to reduce the immediate suffering of individuals. Peacekeeping, peace building and peace enforcement are said to be means of fostering order in countries experiencing various states of chaos and division. The UN's forces designated for such activities leaped from 10,000 in the late 1980s after the Cold War to nearly 80,000 by the mid-1990s. Along with that expansion has come a willingness to 'do something' in a far wider range of situations than had been the case during the Cold War.[4]

The greater enthusiasm for humanitarian operations in some situations has not been met with unanimity of purpose nor necessarily led to actions deemed successful. The examples of UN operations in Somalia, the failure to take significant action to avoid the massacre at Srebrenica, and the slow and arguably counterproductive initial response to the genocide in Rwanda are examples used to suggest the limits of international efforts. The dilemmas and difficulties associated with military interventions are multiple and persistent.[5] As with Srebrenica and Rwanda, action might be deemed too little, too late, or as in the case of Somalia, action can be deemed too overbearing. International interventions into divided societies are unlikely to be seen as impartial by all parties. The credibility of a particular operation must be built over time and is thus always open to re-evaluation based on new developments. Furthermore, interventions to deliver aid might save lives, but often only by supporting warlords or ruining local economies.[6] The desire by some nations to be seen as doing something about specific conflicts has not necessarily been matched by a political willingness to accept the casualties to national soldiers that this might entail. After the televised capture of Western troops in Bosnia and their murder in Somalia, enthusiasm for military intervention waned in many states. Finally, efforts to seek justice and secure peace do not always pull in the same direction.[7]

One of the major areas of contention in humanitarian interventions is the use of force. In response to human atrocities, calls for such actions come from many quarters. When UN, NATO or other such agencies work in areas where there is little consent for their presence, though, the use of force has the potential to provoke much animosity. Even within 'coalitions of the willing', the relationship between international objectives and the application of force can provoke much contention.[8] The ability of peace-enforcement missions to be impartial and not influence the political outcome of conflict is limited.[9] The possibility that, once on the ground, a peacekeeping mission might shift to peace enforcement and then to war-fighting is a major concern from experiences in Somalia.[10] Peacekeepers trained as soldiers do not necessarily have the sort of skills necessary for the situations they face. Force should be proportionate, lawful and accountable, but the meaning of such terms is not straightforward and, when soldiers from different countries mix, considerable scope exists for different evaluations of particular actions.

For the presence of force to deter offensive actions requires that it be seen as credible. Non-lethals are said to help find force means that meet the requirements of peacekeepers and others by enabling a continuum of force responses. Situations of use include crowd control, incapacitating individuals, clearing and denying access to areas, and disabling equipment. In these sorts of circumstance, non-lethals would function as tactical aids in similar ways as mentioned in previous chapters for public-order or routine policing activities.

The 1995 withdrawal of UNOSOM II forces from Somalia in Operation *United Shield* is generally credited as the first major military force deployment involving non-lethal weapons.[11] Here, sticky foams, caltrops, pepper sprays, target illumination devices and beanbags were deployed, though not widely used. Lorenz attributes the latter to the inflexible rules that placed major restrictions on when they were deemed acceptable.[12] Various non-lethal weapons, however, had been employed in other missions, such as during Operation *Restore Hope* in Somalia that started in late 1992. To avoid the 'snatch-and-run' of supplies from assistance convoys and compounds, Western military forces made use of batons and various incapacitant pepper sprays. Tear gas, though available, was not used because of the poor health of the local population and the likelihood for indiscriminate effects.[13]

Some critical commentary has been made about such modest employments of non-lethals. Stanton, for instance, argued that instead of proving the civility and restraint of users, the deployment of such weapons opened up forces to being second-guessed.[14] He maintained that the disturbances encountered in peacekeeping operations are distinct from civilian riots in Western countries. Riots in the former serve as pretexts for, and means of, warfare. Many individuals are armed and intent on causing violence. The greater threat justifies different responses.[15] In these situations, the likelihood of the immediate resort to overwhelming (lethal) force to restore order needs to be present. Rather than stemming from a lack of proper technological options (given that tear gas and kinetic munitions have been around for some time), what has made the use of force in peacekeeping problematic has been a lack of clear policy goals about the ends of intervention and the political will to implement policies.

The lessons of past and recent tactical employments of non-lethals are worthy of detailed examination. Unfortunately, much less exists in the way of detailed accounts of military uses of non-lethal force compared to that for deployments by the police. Partially because of this situation, this chapter does not focus on such deployments. Rather, attention is turned toward the possibilities envisioned by new technologies and how facets of military operations, particularly those associated with humanitarian missions, might alter because of non-lethals. Chapter 3 discussed a number of technologies said to be available in the near future: mobile, highly discriminate lasers that cause temporary paralysis, choking and pain; chemical agents designed to provoke anxiety; calmatives that induce mass sleep; and micro-organisms that degrade materials. Their possible use deserves consideration.

FUTURE SCENARIOS

Let us then move on to consider expectations about the next generation of non-lethals. In an effort to inform funding decisions taken across the American armed forces, in 1998 the US Marines outlined several hypothetical scenarios in its *Joint Concept for Non-Lethal Weapons*.[16] One such setting is that of humanitarian intervention by US forces in a divided and conflict-ridden developing country. The local scene is such: in the middle of armed factions vying for power, aid must be delivered. Several potential situations within this overall conflict are elaborated. For instance, in one, the United States deliberates about the merits of taking some sort of action against a neighboring state that supports one of the parties to the conflict and thereby undermines aid attempts. While as yet not directly confronting US forces, the potential for future hostility requires some action. A pre-emptive strike with anti-sensor and anti-material weapons using remotely-piloted vehicles is launched against the neighbor. Through this action '[t]he potentially hostile force has suffered no personnel casualties but has been rendered operationally immobile and unable to defend itself against further airstrikes, should these prove necessary'.

Another specific situation envisioned within the overall intervention highlights the crowd-control potential of non-lethals as well as the possible implications of *not* deploying them. After setting up routine security roadblocks around the American Embassy, unfriendly locals make tests of its effectiveness. One day, a vehicle speeds toward the blockade and fails to stop. US forces respond by taking fire, killing those inside. Later, it is learnt that the vehicle's brakes were faulty and the passengers were unarmed and therefore probably not of malicious intent. Despite attempts to make the facts of the situation known, a hostile crowd forms around the Embassy in protest. Some people appear armed. Projectiles such as rocks are thrown, followed by fire-bombs and then firearms. In response, US forces return fire with aqueous foam laced with irritants. A remotely-piloted vehicle disperses OC gas on the crowd. When a mob returns after being dispersed the first time, further options are deployed:

> a helicopter appears some distance away, well out of the effective range of small arms. Unknown to the gathering crowd, this helicopter mounts a non-lethal counter-person-nel area-denial system with standoff capability. From over a kilometer away, the helicopter crew directs the weapon at the largest groups of would-be rioters ... The system takes effect, the people immediately flee.

The commander in charge expresses relief at the non-lethal resolution of the situation but warns, through the media, about the ever present possibility for deadly force should it be called for. While having some similarity to previous military operations, the scenarios discussed in the *Joint Concept* are only hypothetical. Here, as elsewhere, many of the claims about the future utility of non-lethals are speculative.

As described in Chapter 4, there are hopes from some quarters that the introduction of new weapons will one day help bring a battlefield where death is unacceptable and uncommon. Less grandiose, abstract and improbable, in 1999 the prestigious Council on Foreign Relations sponsored an independent Task Force of elite corporate, military, academic and other advisors to examine the military-related prospects for non-lethal weapons.[17] Non-lethal weapons were defined as those devices 'designed to disable enemy forces or incapacitate combatants and others without killing them or causing permanent harm'. The report made various recommendations about US defense policy, such as the need for greater funding and co-ordination of research in the armed services. While acknowledging that non-lethals were no panacea, the Task Force maintained that research efforts should bring not just new technologies, but novel ways of conducting warfare or enforcing economic sanctions. Little joined-up thinking about the possible strategic implications of such devices was said to be taking place between the armed services due to their highly secretive conduct.

Writing during the NATO intervention in Kosovo, the Task Force focused much of its attention on the strategic pay-offs of non-lethal weapons in this action. The following quote provides an indication of the range of utilities envisioned:

> Despite weeks of bombing in the spring of 1999, NATO failed to prevent the expulsion of nearly one million Kosovar Albanians. The Task Force does not suggest that nonlethal weapons by themselves could have prevented this tragedy. But consider how nonlethal capabilities could have been used in the early stages of the conflict, as Serbian troops and paramilitary forces began the grisly work of ethnically cleansing Kosovo:
>
> •   NATO could have jammed Serbian TV broadcasts, replacing them with respected independent news sources such as the BBC. In this way, the Serbian public would have been informed about atrocities against Kosovar Albanians. NATO could also have used this channel to air statements designed to mitigate Serbian feelings of victimhood.
>
> •   At the same time, NATO could have used nonexplosive means to turn off the electricity in Belgrade and keep it off, with occasional respites to allow for the reception of NATO television.
>
> •   NATO could also have issued unobtrusive film or video cameras to Kosovars and NATO agents for recording war crimes in Kosovo, along with appropriate means for transmission out of Kosovo. The images could have been used in information campaigns and in war-crime prosecutions.
>
> •   Electromagnetic pulse and radio frequency weapons could have disabled Serbian air-defense and other military electronic systems.
>
> •   Serbian military headquarters and other sensitive buildings might have been rendered temporarily unusable by the precise delivery of revolting smells.
>
> •   Instead of bombing key bridges, thereby blocking commercial shipping on the Danube, NATO could have blocked military traffic over the bridges via repeated, precision deliveries of 'stick'ems' and 'slick'ems'.
>
> •   Meanwhile, NATO blockade efforts could have been

       enhanced by using super-strong cords to entangle ship
       propellers.

- In the event that ground troops had been introduced or major destructive measures taken in consonance with a largely nonlethal campaign, NATO could have launched an extensive campaign of deception, propaganda, and communication warfare.

As indicated by the types of promises above, the definition of non-lethal weapons used by the Task Forces includes a wide range of technologies, from those discussed in this book to 'non-lethal' technologies such as video-recording devices. In general terms, the promise for more weapon-like applications is much the same as that made elsewhere. Whereas conventional weapons such as explosive bombs work by blowing up bridges, in their place, alternative forms of non-lethal weapons could function that resulted in comparatively less destruction and thus a greater legitimacy for force.

SOME INITIAL ISSUES

Let us consider these scenarios in relation to two questions: what capabilities are sought from the technology and what are likely to be key sources of contention? Critical claims about non-lethals surveyed in this book would offer a number of counters and qualifications to the scenarios. Each of these casts doubt on just how realistic the deployments envisaged are. To start with, it is suggested that non-lethals will operate in particular ways, resulting in anticipated effects. 'Stick'ems' stick up bridges, area-denial systems take effect, and revolting smells render areas (temporarily) uninhabitable. Non-lethals are tools that act as means to the ends to which they are put. As in the example given in Chapter 4 of scenarios by the RAND Corporation of dispersing chemical 'knock-out' agents in mega-cities to allow combatants and non-combatants to be sorted through, achieving the promise set out for the technology will require that it functions as advertised, and in some cases only as advertised. This promise has been attributed to non-lethals in other scenarios.[18]

Previous arguments in the book would suggest that, in relation to concrete cases, there is much scope for disagreement regarding effects. In an effort to draw out some of the practical complications of future non-lethal weapons, Robin Coupland of the ICRC commented:

    [Their purpose] is to 'disable'. This sounds better than inflict-

234

ing disability and does not immediately beg the difficult question of how long the person will be disabled for. Will blinding be permanent? Will the various energy forms that target the functions of the central nervous system leave the victim with permanent neurophysiological effect? Can entangling agents asphyxiate? Will a 'calmative' agent only calm? If it is established what energy output or concentration is non-lethal or temporary, you have also discovered what is lethal or permanent ... Rather than sutured wounds, skin grafts, or amputations, will the soldiers who have survived battlefields of the future return home with psychoses, epilepsy, and blindness inflicted by weapons designed to do exactly that? Should not these questions be considered before such weapons are deployed?[19]

Further questions might be raised about the likelihood of achieving specific effects in relation to administering any aftercare necessary for ensuring that weapons are relatively benign. At least for the UN, fairly persistent concerns have been raised about the lack of resources, logistic capability, and personnel in peacekeeping missions,[20] conditions that do not bode well for providing decontamination or other aftercare.

The precision of effects is also married to other capabilities sought. One such implicit characteristic stressed is the importance of maintaining a distance between users and targets.[21] Current research efforts to develop unmanned aerial vehicle non-lethal payload delivery systems perhaps express this desire most fully.[22] The possible implications of distance are multifaceted. As the bombing by NATO of Yugoslavia in 1999 and the US bombing of Afghanistan in 2001/2 showed, high-level flights that keep considerable distance from the ground are capable of bringing about substantial damage without a loss of life for users. When this is the overriding concern, maximizing the gap is the safest way to proceed. Whether or not creating distance is compatible with the selective application of force (especially against mobile objects or persons) seems less certain. Some also contend that physical distance helps create a psychological distance that is key in enabling individuals to kill and injure.[23] In a public-order setting, distance helps minimize the sort of face-to-face confrontations that are likely to increase the adrenaline levels of participants. In circumstances where operators are physically removed from the effects of technology, however, the possibilities for achieving discrete and measured effects are likely to be reduced. Other things being equal, the ability to receive necessary feedback generally becomes more difficult as distance increases. Ensuring a baton causes minimum or proportionate harm is one thing, ensuring the effectiveness of an anti-personnel area-denial system with standoff capability administered from over a kilometer away is another. Actions other than the display and employment of force, such as dialogue between parties (say with a view to de-escalating conflict), also generally become more difficult over significant distances.

Other points about the practicalities of precise effects can be raised. Where a military opponent is expected to be 'temporarily disabled' but still able to attack after being hit, many users may prefer to 'tune' weapons to a high setting to minimize the risks of retaliation. More generally, concerns can be raised about gauging the effects of non-lethals. The Council on Foreign Relations Task Force proposed electromagnetic-pulse and radio-frequency weapons could have disabled Serbian air-defenses. As mentioned previously, while tiny carbon fibers were used in the Gulf War for such purposes, this was followed up by conventional bombing sorties to ensure that the equipment was indeed unworkable.

Many recognize the difficulties associated with achieving precise and desired effects and this in turn has lead to various supplementary innovations. BAe Systems (formerly British Aerospace), for instance, is researching sensor devices that seek to enable the real-time monitoring of the anti-personnel and anti-material effects of non-lethal weapons.[24] Because traditional forms of battle damage assessment, based on the destruction of targets or other visually observable effects, are not appropriate for many non-lethals, efforts are being undertaken to develop detectable and quantifiable measures of effectiveness. The aim is to gauge the thermal signature of equipment so as to be able to determine whether it is capable of functioning. Likewise, the reliance on subjective appraisals of the behavior of crowds is no longer taken as sufficient. Sensing equipment to detect infra-red wavelengths is being developed that would give some measure of the physiological response of individuals targeted with certain weapons (say, in terms of their perspiration levels). The aim is that crowds will be monitored in real time for their reactions so security forces can respond accordingly. How data related to such indicators can inform an understanding of group dynamics has not yet been elaborated.

The previous paragraphs suggest likely areas of dispute about the feasibility and implications of non-lethals should they be employed as outlined in the aforementioned scenarios. Such concerns, though, are just the start of asking how debates about the merits of non-lethals might unfold. It is generally recognized that enhancing the legitimacy of recourse to force requires more than that a given weapon can have the desired effect, but that this is interpreted as justified. Whereas BAe Systems seeks to monitor effects, Dutch researchers are devising an

---

FIGURE 10.1
CHECKLIST CRITERIA FOR THE OPERATIONAL UTILITY
OF NON-LETHAL WEAPONS

- Does the NLW have effect on the specified target? What is the nature of the effects?

- Is the NLW accurate enough to have any effects on the target(s) in this scenario?

- Are the effects completely understood? Can physiological remaining effects occur to the user or targets of the NLW? What is the probability of occurrence?

- Will escalation occur because the NLW invites us to use it whenever and wherever we want?

- An NLW is never merely a replacement of an existing weapon! Most NLWs can be regarded as complementary to a lethal weapon and therefore should be deployed in conjunction. Is the NLW a strong complementary (additional) to existing lethal weapons? Should one reconsider this in the RoEs'?

---

operational support 'Frame of Reference' database to assess a wide range of issues associated with the acceptability and usability of non-lethals.[25] Answers are sought to various questions, such as 'are the troops sufficiently trained to deploy the NLW [non-lethal weapon] system?', 'are there any conventions or treaties applicable?', 'will the effect derived by the NLW conflict with the local social/cultural back-ground?', and 'will the deployment of the NLW influence the public opinion or acceptance with respect to use when exposed in the media of the local or own society?'.[26] The short-term aim of this activity is to develop a checklist for evaluating operational utility. Figure 10.1 lists a few of the questions asked for this purpose.[27] The ultimate aim of the 'Frame of Reference' initiative is to establish quantifiable measures of the overall utility of a weapon by getting scientific and military experts to give answers to the questions mentioned for a given weapon in a particular scenario. Through a complex mathematical weighting of all the criteria, overall measures of utility will be derived.

Such a checklist and the other 'Frame of Reference' questions high-light a number of issues associated with non-lethal weapons identified in previous chapters. The argument of this book, though, would suggest that the process of answering such questions and what any particular determinations should then imply are less than straightforward. Leaving aside just how different criteria might be weighed, the answers to the questions are often disputed because of the lack of information, the conditionality of claims, and alternative interpretations of evidence. Escalation is not an inherent property of a given weapon, but is the result of the interrelation of technology and humans, where beliefs held

237

about the potential for escalation are key in whether this takes place. Furthermore, determinations about the accuracy of weapons or the adequacy of the training of troops might be taken as given in many circumstances. When something is seen as going wrong and responsibility is sought, however, those givens may well become open to qualification. Whether particular weapons will influence public opinion when exposed in the media depends on just how and what exposure is given. In other words, determinations of these issues should be regarded as products of debates about the weapons as much as they are inputs. It is not just that there are different views about the likely merits of given technologies, but that alternative interpretations and ways of attributing effects to technology exist.

REVISITING NON-LETHALS AND KOSOVO

Let us consider some of the points so far in relation to the Kosovo crisis and the previous speculation about what roles non-lethals might have played therein. This section does not attempt to determine what response should have been made in either 1999 or prior to that,[28] but considers something of how the debate about the merits of the use of force in this intervention unfolded, what were some of the main areas of debate regarding force, and how non-lethal capabilities might have figured in these. As in other interventions, in this one questions were raised about the aims and effects of weapons, the scope or different interpretations of weapons, and the aspects of a conflict relevant for consideration.

On 24 March 1999, US-led NATO forces began the bombing of the Federal Republic of Yugoslavia with the goal of stopping human-rights atrocities from taking place in Kosovo. As suggested in the recommendation given by the Council on Foreign Relations above, had non-lethals been developed, they could have done 'much to legitimize the NATO operations in Kosovo and Serbia, avoiding some of the adverse reaction there and in Russia'.[29]

While many military leaders and political pundits generally present the outcome of the war as a great success, this evaluation is not universal. Tracing out something of the reasons for this can assist in the understanding of the likely scope for contention with non-lethals in future operations. Throughout the ten weeks of bombing in Operation *Allied Force*, the campaign was said to strive to cause minimum harm to civilians. The problem for NATO was not with the Serbian people but with the then President, Slobodan Milošević, and his security agencies responsible for atrocities against ethnic Albanians in Kosovo. Accepting this humanitarian intent, then at least some concerns can be raised about whether ends and means were at odds. The extent of Yugoslav civilian casualties has been a major source of concern for some. Human Rights Watch claims to have verified 500 deaths to Yugoslav civilians, a number higher than NATO estimates but lower than official Yugoslav statements.[30] More Serb civilian casualties were reported in the first three weeks than those to all sides in the three months prior to the start of the NATO bombing. It was reported that by August 1999, as many as 164,000 of 200,000 Serbs living in Kosovo prior to the bombing had fled.[31] For proponents of non-lethal weapons, any disparity that existed between means and ends in the intervention can be taken as supporting the availability of new means. Instead of casting doubt on the wisdom of the operation, non-lethals should be brought in to mitigate the resulting loss of life on all sides.

Whether or not this is deemed realistic, other commentators have drawn on issues about the intervention to question its ends. A basic criticism has been raised as to whether the war was conducted to save Kosovo Albanians or to save the credibility of NATO as the premier security agency in Europe. Here, why any action was taken mainly derives from institutional priorities and geopolitics, where the rhetoric of humanitarianism was selectively drawn on to justify certain actions.[32] Johnstone, for instance, argued that the war had other objectives than saving innocent lives, such as: asserting US authority on what constituted a humanitarian cause, instilling a lasting US presence in the region, and reaffirming the legitimacy of warfare (now labeled humanitarianism).[33] The framing of the conflict as one between evil Serbian perpetrators and innocent Albanian civilians was a selective and simplistic story about good versus evil. Attempts to resolve the conflict short of war were said to be half-hearted or possibly disingenuous. She argued, for instance, that the Organization for Security and Co-operation in Europe mission to verify ceasefire arrangements prior to bombing probably prepared the way for war and helped ingrain aggression on all sides. Others have voiced related doubts about the necessity for the aerial assault. Early talks in Rambouillet about peace in the area have been presented as motivated by a desire to provide clear justification for the use of force by presenting an agreement unacceptable to the Serbs, rather than by a desire to find a negotiated peace.[34]

Much of the dispute about specific instances of the use of force in this operation focused on what were 'legitimate designated military targets'. The international laws of warfare forbid attacking civilian infrastructure. As the bombing of a narrow range of military targets in the earlier days proved ineffective to bring about the desired settlement, the matter of just how far the range of NATO's targets should be extended came to the fore. Particular concerns were raised about the appropriateness of strikes against Serb Radio and Television in Belgrade, New Belgrade heating plant and several bridges. The scenarios of the Council on Foreign Relations maintained that these sites were legitimate military targets, but suggested that non-lethals might have produced the desired consequences without the same number of deaths. Arguably, in this scenario a rather crude psychology is assumed, where those on the ground would have taken the lack of outright physical destruction of these objects as an indicator of NATO's desire to inflict minimum damage. Implicit is the assumption that non-combatants would agree that conventional weapons would have been the only, and an otherwise appropriate, alternative.

It seems obvious that those in countries such as Romania, Bulgaria and Ukraine would have preferred if the Danube river had not been blocked for months by the physical destruction of bridges in Yugoslavia. Just how long and in what way 'stick'ems' and 'slick'ems' could have rendered the bridges unusable is a matter for consideration. Whether or not such technology would have provided a viable alternative depends on the basic motivation attributed to the bombing. That the destruction of such infrastructure happened at all is taken for some as an indication of a striking indifference to the needs of the region and the desire for pure destructive spectacle.[35] That many bridges were bombed during the daytime, and that such acts brought civilian deaths (for instance, as in Varvarin), has been portrayed as part of a psychological warfare strategy;[36] one designed to break civil resolve with scant regard for possible casualties. What role non-lethals, such as revolting smell devices, might have played as intimidating tools as part of psychological warfare techniques is not touched on by the Council's Task Force, but this issue would seem to merit attention in light of previous experiences in places such as Vietnam. A danger with non-lethals, as suggested in previous chapters, is the possibility that they might legitimate targeting a wider range of locations given their supposed benign effects.

Additional sources of division exist about the wisdom of the bombing. Take, for instance, the mass exodus of ethnic Albanians from Kosovo. The Task Force argued that '[d]espite weeks of bombing in the spring of 1999, NATO failed to prevent the expulsion of nearly one million Kosovar Albanians'. That expulsion took place along with 'cleansing' operations; where an estimated 10,000 Kosovar Albanians were killed. Non-lethal capabilities were said to help to counter such outcomes. Others would suggest that the timing implied in such statements is faulty; the bombing came before the majority of atrocities rather than after them. Instead of preventing the expulsion, the NATO bombing set the groundwork for it. When regular Yugoslav forces and

Serb police were struck and unable to repel the air attacks, and with NATO ground forces ruled out from the start, that the security forces turned toward those on the ground was said to be hardly surprising.[37] US-NATO commanding officer for the operation, General Wesley Clark, reportedly commented that it was 'entirely predictable' that Serb atrocities would intensify after the NATO campaign began.[38] The ethnic conflict that had been taking place for years escalated into all-out war with the dropping of bombs. In other words, the intervention precipitated the catastrophe it was meant to prevent – the flight of Kosovars, whether that was to the south or north.[39]

In short, then, with regard to Kosovo there were disputes about effects, about the appropriateness of means, and about the ultimate ends. Of course, it is hardly surprising that disagreement exists about the agenda and the outcomes of military interventions. With a focus on the possible contribution that non-lethals might have made, there are important questions about what follows.

The previous paragraphs suggest the importance of the presentation of the necessity and effectiveness of the use of force, as well as the knowledge held about such incidents. Taking the effects and legitimacy of technology as the outcome of convincing interpretations, the media assumes a significant role in public disputes about non-lethals. It is worth commenting on the place of the mainstream media in more detail. Just what role news organizations played and how they influenced debate are matters of contestation where differing attributions of intent, truth and effects are offered.

As previously stated, fears have been expressed by those in the military and elsewhere that a hypersensitive public has made the forces of Western democracies quite risk-adverse. The abilities of the armed services to wage war, or the police to maintain order, are being undermined by a 'squeamishness' about killing or causing damage. That images and accounts of civilian deaths (so-called 'collateral damage') from NATO attacks were televised around the world without corresponding images of Kosovar Albanians being killed was said by the Council on Foreign Relations' Task Force to be a 'propaganda debacle'. The overall need for sustained effort in ensuring that force is presented as appropriate and justified has become a topic of fairly widespread commentary in military circles. The so-called 'CNN effect', the global broadcast of live news from the sites of coverage, demands a new set of responses compared to past military-media forms of relations. As Stech argued, live on-the-scene images require accompanying stories to ensure that the proper interpretations are made.[40] From this, those in the military need thought-out strategies for presenting themselves and interacting with the media who might otherwise be antagonistic.

Extending Stech's line of thinking to the presentation of the operation of non-lethals, possible contentions include all those matters of ambivalence noted so far:

> Media coverage might elicit such negative public or political reactions as, on one hand, that NLWs violate international treaties, damage the environment, make war more likely by reducing the destructive consequences, maim and injure noncombatants, cost too much, or simply do not work; or on the other that NLWs reflect a sentimental or naive view of war and a lack of resolve to defend national interests, that such weapons risk the lives of soldiers, compromise operational effectiveness, are insufficiently potent to punish aggressors, and are 'politically correct' but militarily irrelevant.[41]

In terms of Stech's recommendations to supply a proper contextualizing story, past chapters would suggest that one such framing is likely to be that non-lethals reduce numbers of injuries compared to those from conventional options.

Elsewhere, the agenda of the mainstream Western media in covering the bombing has come under much criticism as being too aligned with that of NATO. Points of condemnation include: the media's orientation in favor of mobilizing support for the intervention rather than bringing out the complexities of the conflict and the moral standing of many participants;[42] the lack of examination of possible alternative motives, hidden agendas, or inconvenient facts;[43] and the controls placed on the access.[44] With regard to the effectiveness of the bombing, a major source of contention after the campaign was the number of objects hit by airstrikes. While NATO representatives initially claimed that the Serb military forces were 'seriously damaged' and Western media sources sometimes (perhaps often) reproduced these statements without qualification,[45] after the bombing, estimates were significantly revised downward.[46] Seemingly the most significant damage was done to fixed targets such as 'dual-use' infrastructure. Depending on one's orientation, the disparity between 'during' and 'after' figures might be taken as an indication of media manipulation by the military or a reflection of the difficulties of gauging battlefield effects. Assuming the former, the suggestion that the mainstream media should be dependent on such sources for authoritative and credible information, and generally be unquestioning of them, casts doubt on their rigor.

There are thus alternative ways of making sense of media organizations. These suggest different orientations to matters about just whether and why non-lethals might further legitimate interventions. Assessments about the involvement of the media turn on evaluations made about credibility and motivations.[47] The manner in which the legitimacy of the non-lethals depends on their public presentation has already been discussed in Chapter 4 regarding the visibility of the effects. There, alternative appeals were made to 'the reality' versus 'the appearance' of the effects of non-lethals. So, the weapons were portrayed as both sounding and looking innocuous while being quite dangerous in reality, as well as appearing threatening but actually being relatively harmless.

THE FUTURE OF NON-LETHALS AS HUMANITARIAN TOOLS

The previous sections considered how non-lethals might alter conflict, likely areas of debate, as well as what sort of measures are thought necessary to ensure they are interpreted in the 'correct' manner. There are a number of implications that follow for thinking about the future of non-lethals in humanitarian interventions. As with other aspects of the technologies discussed in this book, the scope for debate suggests the importance of considering how notions about what these weapons can and cannot do are established. Stories about the operation of non-lethal weapons are attempts to set out ethical relations between technology and individuals. In contrast to contentions that the use of non-lethals will ensure a greater legitimacy to force because their employment indicates the intent of users to reduce injury, this analysis has suggested that just what non-lethals are taken for will be part and parcel of negotiations about a wide range of issues regarding the appropriateness of force. To ask if the weapons will influence public opinion when exposed in the media fails to acknowledge how media portrayals are themselves managed and how assessments of weapons are bound with assessments of the role of the media. The importance of managing debate to foster certain public perceptions is acknowledged in many discussions within the military about force. As a joint US/UK seminar about the future of non-lethal weapons concluded:

> A robust and effective public affairs campaign addressing-both domestic and international audiences was generally recognized as an essential dimension of the successful use of NLW. The broad goal of such an effort, of course, is enhancing the 'acceptability' of NLW use and, regarding certain types of operations, the acceptability of the use of military force.[48]

This, then, is not just to agree that the legitimacy of these weapons depends on, say, the management of media presentations. Rather, this analysis has suggested that the possibility that non-lethal weapons will help resolve disputes is limited because what they are taken for is a matter disputed and disputable. Along these lines, past chapters considered various ways in which complex situations were presented in order to offer an account of the 'actual' effects of weapons and, by extension, the merits of their deployment. Much of this has turned on questions about how to speak about the relation between the 'social' (or 'non-technical') and the 'technical'. Depending on how one understands and defines these terms, varying assessments can be made. For instance, to the extent that one takes the effects of technology as known, predictable, and relatively acceptable, then when problems arise, attention is cast to various 'social' issues, such as the circumstances in question or training provisions. The inability of defense forces in Israel, armed with non-lethal weapons, to respond to Palestinian civil disturbances without inflicting civilian deaths might be taken as a failure of equipment capabilities, a lack of concern about causing deaths, or some other such determination. The question of how the 'social' and the 'technical' are understood extends to wider issues about the promise on offer. In debates about the use of force, such as the future importance of new weapons for the conduct of military operations, the devices in question can be seen as offering novel technical options, or their promise can be seen more as a 'socially engineered product of impression management than of specific technological advances'.[49]

The matter of how praise and blame is attributed in discussions about the use of force – whether this rests with the properties of the technology itself; social considerations about how it is used, by whom, to what end; or some combination of the two – is particularly pertinent in relation to non-lethal technologies because of the alternative and contradictory ways these devices are described. On the one hand, non-lethals are said to be tools just like any other. They are means to ends that can be good or bad depending on the intent and competencies of users. On the other hand, as has been mentioned throughout this book and in the scenarios above, such weapons are also presented as embodying a certain intent, or political values (progressive or repressive), within their design. It is recognized that these matters are disputed. So the Council on Foreign Relations acknowledges the scope for alternative interpretations about non-lethals by NATO, but this in terms of whether one properly acknowledges the real intent of the technology. That the supply of electricity could be cut off by non-lethal means, save for the transmission of Western news programs, was seen as deserving some level of acceptance by those being affected as well as by domestic audiences.

Adopting 'analytical skepticism', this book has treated attributions made of technology as just that, attributions which rest on particular ways of characterizing the technical and non-technical. To frame debates about them in terms of whether it is a matter of social impression management or due to the actual technical capabilities is to ignore the way such a discussion is itself already a contingent way of framing the issues at stake, as well as the work that goes on in dividing the 'merely social' from the 'merely technical'. Definitive statements about the operational effects of specific weapons assume a particular understanding of the technology and the situation. These are always open to question. As elaborated in earlier chapters, what is taken as the proper context for evaluation should be understood as a resource for supporting particular interpretations. As different contexts are drawn on (for instance, a highly critical or complacent media), alternative stories are given of the operation of weapons. There is little hope in this situation that *a* 'frame of reference' could provide the proper viewpoint for indicating the suitability of such weapons.

Non-Lethals as Justifying Force?

Take these points in relation to concerns about the potential for non-lethal weapons to lead to a growing resort to force. A consideration of this topic provides a small case for considering the possible terms of future debates about the strategic implications of non-lethal weapons; a case that incorporates issues discussed throughout this chapter. The potential escalatory (or de-escalatory) role of non-lethals has been a subject of consideration in previous chapters, especially at the tactical level. As we have seen, the possible strategic or operational dangers center on whether the weapons are relatively 'low-cost' options that might encourage the resort to force. It is possible to see the potential for this expressed in the scenarios above. The US Marines contended:

> Because we can employ non-lethal weapons at a lower threshold of danger, commanders can respond to an evolving threat situation more rapidly. This allows US forces to retain the initiative and reduce their own vulnerability. Thus, a robust non-lethal capability will assist in bringing into balance the conflicting requirements of mission accomplishment, force protection, and safety of noncombatants. It will therefore *enhance the utility and relevance of military force* as a US policy option in an increasingly complex and chaotic international environment [emphasis mine].[50]

One pertinent question in relation to disputes about force is whether the goal of increasing the relevance of military force as an instrument of policy should be pursued, even when conducted by non-lethal means.

Just whether and how the availability of non-lethals might encourage treating situations as resolvable through force is a key issue in assessing their contribution to humanitarian interventions. In thinking about the future of debates, let us consider lines of argument that might form around questions about the merits of the recourse to force in the future in such operations. As elsewhere, questions can be asked about the likelihood that non-lethals will function as presented. The Marines' pre-emptive strike with anti-sensor and anti-material weapons was said to render the armed forces of a potentially hostile group operationally immobile without casualties. Assuming that this can be verified, what happens from there? What if these forces then become hostile in some manner? What counter-counter-response should follow?

With regard to contentions that the availability of non-lethals might encourage a growing resort to force as a means of policy, while force may be appropriate in some situations, the worry is that commitments might be made to use force in situations that are dangerous or inappropriate. Here, following the lines of argument described in previous chapters, the functioning of these weapons might *transform* the options and likely actions of users and organizations. A line of opposition to such claims is that technologies in themselves don't cause any change in behavior, but merely *add* to the range of options. Where force is deployed inappropriately, that it is the result of poor decisions taken by people in policy positions.[51] How can mere technology be blamed for promoting the use of force? Such a counter line of argument has its own counter. Uncertainty, uncontrollability, unpredictability are key themes in the history of weapons. Here, the simple logic of means and ends is of limited value in understanding how weapons might alter the character of conflict. 'Intent' – whatever it is taken as – is thus too precarious a determiner of the implications of technology in practice given the types of complexities and uncertainties of conflict. As outlined previously, the introduction of weapons into warfare is rife with ill-fated predictions. So, the issues at stake are more diverse than simply assuming that non-lethals will operate according to decisions made about them.

This book has sought to consider a wider range of issues than whether there are certain practicalities that might hinder the functioning of non-lethals, or that their use is highly complicated. Asking whether their availability will lead to a greater recourse to force because of the humane status of the weapons, or whether non-lethals should be understood as tools whose merits depend on the ends to which they are put, already frames the issues at stake in a contingent manner that implicates these weapons in action and suggests ways of understanding what the key issues are. Let us consider this with regard to contrasting claims about the additive or transformative effects of non-lethals.

On the one hand, to say that the humanitarian ends of the employment of technology should be understood as determined by decisions made by certain individuals is to suggest that there is some given technology with known effects across varied circumstances that can be evaluated in terms of the actual ends to which it is put. Of course, in this treatment of the issues, unintended consequences might be noted, but these are treated as side effects owning to some contingency. Once these parameters are established, debates center around whether means and ends are aligned and the 'real' ends sought. These were the terms that characterized much of the discussion above about NATO's use of force in Yugoslavia. So, did NATO really bomb bridges during the daytime to intimidate civilians? To what extent were its goals really humanitarian? As suggested in Chapter 8, though, just how responsibility is allocated and how technologies are defined should be seen as products of the handling of uncertainties and indeterminacies. To say that the appropriateness of force options rests with decisions made about them by users or policy-makers is to frame the issues at stake in terms of whether or not the intent of decisions is appropriate and realized rather than drawing attention to the efforts undertaken to promote particular understandings of intent. Readings of intent are the product of various lines of presentation, where the role of the media is arguably of considerable importance in influencing public discussions. Ignoring such issues provides a blinkered view of the range of issues at stake in debates about non-lethals. Instead of politics being constituted throughout the process of describing, designing and deploying weapons, a focus on the choice of certain decision-makers has the value of technology simply deriving from deployment decisions.

In contrast, to suggest that non-lethals might transform the actions by justifying and therefore encouraging the greater resort to force is to treat the technology as having a certain capacity built in. With this as the focus, attention turns away from seeing claims about what weapons can and cannot do as contingent attributions negotiated during the entire course of their development and deployment. A built-in status is problematic on a number of grounds. In the first regard, what counts as 'humanitarianism' (or its opposite) is a matter much disputed. Further, there are worrying implications of treating intent as built-in (whether this is for good or for bad). Just as claims that a certain number of deaths are inevitable from the use of any weapon might deflect attention away from issues about how weapons were employed in specific circumstances, so, too, justifications that the availability of certain options compelled a certain response to a situation should be questioned.

As always, the reason for certain outcomes is potentially or actually a topic of controversy. Future debates about non-lethals are likely to be characterized by attempts to define the 'social' and 'technical', as well as the move between them to offer a particular account of events and the merits of this weaponry. Just how such conventions are used in the future deserves close scrutiny.

FINAL REMARKS

In conclusion, to the extent that non-lethals are deployed in future operations, there will be various questions asked about the necessity of this, what effects the weapons could have had, and what was done. This analysis would suggest that the possibility that non-lethals will resolve disputes about the appropriateness of force are much more limited than sometimes claimed. Multiple interpretations of appropriateness of force stem from alternative assessments of what the relevant factors for consideration are and how these are understood.

In thinking about how the further introduction of non-lethal weapons into the operational planning and conduct of military missions might change future debates about the merits of force, there are likely to be ever-increasing efforts by those wishing to comment on the merits of force to define the proper framework for evaluating these technologies. Non-lethal weapons will not ensure certain propaganda or public-relations results; their evaluation will very much be the objective of such activities. Contentions that these weapons are simply means to ends, or embody a humanitarian status, fail to note that such views are themselves particular products of portraying the key concerns. In thinking about the future role of non-lethals in humanitarian missions, this chapter has suggested that the sorts of dichotomy discussed in Chapter 4 about the threats and promises of non-lethals – whether they are escalatory or not, or whether they allow for precise targeting or not – are themselves ways of questioning already implicated with assumptions about the functioning of weapons. The future of non-lethals will be one intricately bound with attempts to control the interpretation of events and what counts as a proper frame of reference. In this, what technologies do and do not do, as well as what they are for, will be part and parcel of debates. Thus, there seems little reason to assume that the availability of new forms of non-lethals will decisively resolve assessments about the legitimacy of force.

NOTES

1. See as well Barry, Major General J., 'Beyond the Rubber Bullet' (2002). See http://www. dtic.mil/ndia/2002nonlethdef/Barry.pdf
2. For general discussions of peacekeeping see Ramsbotham, O. and Woodhouse, T. *Encyclopedia of International Peacekeeping Operations* (Santa Barbara, CA: ABC-CLIO, 1999); Kelly, M., *Restoring and Maintaining Order in Complex Peace Operations* (The Hague: Kluwer, 1999); and Jocelyn, C., *Soldiers of Diplomacy* (Toronto: University of Toronto Press, 1998). For a discussion about duplicity in what areas are worthy of intervention see Chomsky, N., *The New Military Humanism* (London: Pluto, 1999).
3. Kaldor, M., *New and Old Wars* (Cambridge: Polity, 2001).
4. Whitman, J. (ed.), *Peacekeeping and UN Agencies* (London: Frank Cass, 1999).
5. Adebajo, A. and Sriran, C., *Managing Armed Conflicts in the 21st Century* (London: Frank Cass, 2001).
6. Malone, D. and Wermester, K., 'Boom and Bust?', *International Peacekeeping* 7, 4 (2000), 37–54.
7. For a discussion of these issues see Shawcross, W., *Deliver Us from Evil* (London: Bloomsbury, 2000).
8. Sanderson, J., 'The Incalculable Dynamic of Using Force', in Biermann, W. and Vadset, M., *UN Peacekeeping in Trouble* (Aldershot: Ashgate, 1988).
9. Berdal, M., 'Lessons not Learned', *International Peacekeeping* 7, 4 (2000), 55–74.
10. Rose, M., 'Military Aspects of Peacekeeping', in Biermann, W. and Vadset, M., *UN Peacekeeping in Trouble* (Aldershot: Ashgate, 1988).
11. See Wallace, V., 'Non-Lethal Weapons', in N. Lewer (ed.), *The Future of Non-Lethal Weapons* (London: Frank Cass, 2002).
12. Lorenz, F., 'Non-Lethal Force', *Parameters*, Autumn (1996), 52–62.
13. Lorenz, F., 'Confronting Thievery in Somalia', *Military Review* (1994), 46–55.
14. Stanton, M., 'What Price Sticky Foam?', *Parameters*, Autumn (1996), 63–8.
15. See as well Peters, R., 'The New Warrior Class', *Parameters*, 27, 6 (1996), 66–7.
16. Marine Corps, *Joint Concept for Non-Lethal Weapons* (Quantico, VA, 1998).
17. Council on Foreign Relations, Independent Task Force, *Non-Lethal Technologies* (Washington, DC: Council on Foreign Relations, 1999).
18. See Alexander, J., *Future War* (New York: St Martin's Press, 1999) and Morehouse, D., *Nonlethal Weapons* (Westport, CT: Praeger, 1996).
19. Coupland, R., ' "Non-Lethal" Weapons', *British Medical Journal*, 315 (1997), 72.
20. Shawcross, *Deliver Us from Evil.*
21. See as well UK Steering Group for Patten Report Recommendations 69 and 70 Relating to Public Order Equipment, *A Research Programme into Alternative Policing Approaches Towards the Management of Conflict* (Belfast: Northern Ireland Office, 2001) and Lewer, N. and Schofield, S., 'Non-Lethal Weapons for UN Military Operations', *International Peacekeeping*, 4, 3 (1997), 71–93.
22. See e.g. Abaie, M., 'Unmanned Aerial Vehicle (UAV) Non-Lethal(NL) Payload Delivery System', presented at Non-Lethal Defense III, 25–26 February 1998. See www.dtic.mil/ ndia/NLD3/aba.pdf
23. Grossman, D., *On Killing* (London: Little, Brown and Company, 1995).
24. Naraidoo, M., 'What is the Equivalent of Battle Damage Assessment for Non-Lethal Weapons', in *Non-Lethal Weapons*, Proceedings of the 1st European Symposium on Non-lethal Weapons, 25–26 September 2001, Ettlingen, Germany (Postfach: ICT, 2001).
25. Delmee, M., 'Frame of Reference and Evaluation of the Operational Value of NLW', in *Non-Lethal Weapons,* Proceedings of the 1st European Symposium on Non-lethal Weapons, 25–26 September 2001, Ettlingen, Germany (Postfach: ICT, 2001).
26. Ibid., 28-5-28-8.
27. Collated from ibid., 28-12-28-15.
28. For a past suggestion of possibilities prior to the start of conflict in 1999 see Transnational Foundation for Peace and Future Research, *Preventing a War in Kosovo* (Lund: TFF, 1992).
29. Council on Foreign Relations, *Non-Lethal Technologies.*
30. Human Rights Watch, *Civilian Deaths in the NATO Air Campaign* (New York: HRW, 2000).

31. Petras, J., 'NATO in Kosovo', *Z Magazine*, 1999.
32. Albert, M., 'Mother Jones, Todd Gitlin & Kosovo', *Z Magazine*. See www.zmag.org
33. Johnstone, D., 'Nato and New World Order', in P. Hammond and E. Herman (eds), *The Media and the Kosovo War* (London: Pluto, 2000) and Johnstone, D., 'Humanitarian War', in Ali, T., *Masters of the Universe?* (London: Verso, 2000).
34. Little, A., *Moral Combat: NATO at War*, aired on BBC2, 12 March. See news.bbc.co.uk/hi/english/world/europe/newsid_674000/674056.stm
35. Ascherson, N., 'Damning the Danube', *The Guardian*, 24 November 1999.
36. As in Human Rights Watch, *Civilian Deaths in the NATO Air Campaign*.
37. See Kaldor, *New and Old Wars*, Afterword.
38. See Chomsky, *The New Military Humanism*.
39. Callinicos, A., 'The Ideology of Humanitarian Intervention', in Ali, T., *Masters of the Universe?* (London: Verso, 2000).
40. Stech, F., 'Winning CNN Wars', *Parameters*, Autumn (1994), 37–56.
41. Coppernoll, M., 'The Nonlethal Weapons Debate', Naval War College Review (1998). See www.nwc.navy.mil/press/Review/1999/spring/art5-SP9.htm
42. Ackerman, S. and Naureckas, J., 'Following Washington's Script', in P. Hammond and E. Herman (eds), *The Media and the Kosovo War* (London: Pluto, 2000) and Hammond, P., 'Reporting Kosovo: Journalism vs. Propaganda', *Transitions Online*. See www.transitions-online.org.
43. Herman, E. and Peterson, D., 'CNN: Selling NATO's War Globally', in P. Hammond and E. Herman (eds), *The Media and the Kosovo War* (London: Pluto, 2000).
44. Keeble, R., 'New Militarism and the Manufacture of Warfare', in P. Hammond and E. Herman (eds), *The Media and the Kosovo War* (London: Pluto, 2000).
45. BBC News, 'Serb Military "Seriously Damaged"', *BBC News*, 26 March 1999.
46. See Ministry of Defence, *Lessons from the Crisis* (London: HMSO, 2000) and BBC, 'NATO Defends Kosovo Campaign', *BBC News*, 16 September 1999.
47. For an illustration of this see Porter, H., 'For the Media, the War Goes On', *The Observer*, 4 July 1999.
48. See US/UK NLW Urban Operations Wargaming Program, *US/UK Non-Lethal Weapons/Urab Operations Executive Seminar – Assessment Report*: 7.
49. Broughton, John, 'The Bomb's Eye-View', in S. Aronowitz, B. Martinsons and M. Menser (eds), *Technoscience and Cyberculture*, (London: Routledge, 1996).
50. Marine Corps, *Joint Concept for Non-Lethal Weapons*.
51. Alexander, *Future War*. The sentiments expressed here are quite similar to those in the phrase 'guns don't kill people, people do'.

11

Conclusions and Recommendations

This examination suggests that the major effort in biological and chemical warfare should be directed toward developing temporarily incapacitating agents which could be defended as being relatively humane – a limited capability for limited wars and appropriate for actions among friendly populations. Concerted studies are needed on techniques to encourage a favorable public reaction to temporarily incapacitating agents.

If such an [nonlethal] incapacitating agent can be developed, we suggest that it not be called BW [Biological Weapon] or CW [Chemical Weapon], but that it be designated by a new term emphasizing our objective of using the most humane agents adapted to the need. Many emotional involvements have grown up around the concepts of chemical and bacteriological warfare, stemming from sources as diverse as World War I propaganda and man's long fight against disease. It would be well not to load this new agent down with such old freight. If such humane agents can be developed, they should be used for the first time in a context which makes it clear that they are alternatives to worse means for gaining desirable goals; if they are so introduced, their reception may well be favorable ...

It was argued that the ideal incapacitating agent should not be classed with the toxic biological or chemical agents and that it should be characterized by some new term, such as 'reinforced tear gas', or 'super tear gas', to emphasize its relatively innocuous nature. Thus, it would not fall automatically under the present political restrictions affecting BW and CW weapons, and it might be accepted as an extension of the basic philosophy of riot control to a larger sphere.

So concluded the US Defense Science Board Task Group on Biological and Chemical Weapons Development in a symposium during 1959.[1]

The findings of the Task Force set the policy stage for work in the 1960s on CS gas in Vietnam and psychotropic drug experiments. The quote expresses something of the tribulations associated with the introduction of non-lethal weapons and the need to plan a response to challenges about their legitimacy. The definitions given to weapons, their classification as relatively humane agents, and the policing of boundaries about their acceptability are all matters seen as requiring proactive management. Support must be mobilized while potential criticisms are deflected. In this the technologies in question, their context, and the criteria for making deployment decisions are acknowledged as sources of future contention.

Following out potential and actual conflicts, this book has sought to understand disputes about the legitimacy of security agencies' recourse to non-lethal weapons. In doing so it has started from and reaffirmed the ambivalence of the resort to such weapons. If their evaluation is framed in terms of whether non-lethals might reduce numbers of injuries and deaths compared to those from other weapons, then it seems unlikely that the answer could be no. If, instead, the key questions asked are whether they might prove lethal or be abused, then again it seems unlikely that the answer could be no. At such levels of abstraction, though, the level of insight gained is limited. This book has sought to go beyond such broad positions in order to ask how the promises and threats attributed to non-lethals are established. In this, a heady mix of appeals to practical necessity, power, idealism and morality are made.

Much of the current literature about non-lethal weapons follows a highly predictable pattern: the need for such capabilities is outlined; descriptions are given of different devices; scenarios for their use are envisioned; and (to varying degrees) caveats are noted. Subject to ironing out a few technical details or factoring in various social, ethical and legal concerns – perhaps requiring the commissioning of expensive expert studies – non-lethals are concluded as having much potential.[2] While not devoid of merit, what such analyses often fail to address are the contentions about meaning, negotiation and disagreement regarding the use of force. It is assumed that the effects are as advertised on manufacturer's brochures, that the devices can be and are used as recommended, and that their legality can be determined by applying agreed international standards. Once these and other aspects are accepted as given, deciding on the appropriateness of these weapons largely becomes a matter of settling on the motivations of those using them. To take this sort of approach, though, is to pass over just how interpretations of non-lethals and their legitimacy (or illegitimacy) are sustained.

As suggested here, particular claims about the legitimacy of the use of non-lethal weapons are often tension-ridden and highly disputed. The reasons for this are multiple and relate to the dilemmas associated with describing and categorizing disputed events.

Determining what particular weapons do and do not do, as well as their effects, requires assessing the credibility of competing claims, the criteria for assessments employed, and the relevance of past experience. In an effort to consider the basis of facts, I have traced the arguments and evidence given in supportive and condemning evaluations. Claims about the safety of a particular weapon can travel through security, legal, medical and public settings. In doing so, claims either become treated as 'facts' or do not. What the facts of the matters are, though, is open to challenge. Decisions about whether a death in police custody was due to restraint techniques or to a chemical incapacitant spray change as medical understanding alters. In other words, what a device is taken for and what it can do are interpreted and reinterpreted over the period of its development and deployment.

Likewise, the appropriate 'context' for understanding the operation of non-lethals is a matter of dispute. The situational context presented for the deployment of weapons – whether that is the policing of public demonstrations, responding to suicide attempts, or enforcing order during humanitarian military operations – is key in assessing the appropriateness of force. A force deemed acceptable in military interventions, even peacekeeping missions, might be wholly unacceptable in domestic policing. Choices made about the types of description offered are important not just in characterizing the events and level of insecurity experienced, but in what counts as reasonable action and the acceptable effects of weapons. Many similar points could be made about contexts for evaluations. As argued, much depends on whether a rather narrow attention is given to the physical effects of weapons in particular cases or whether a broader approach is adopted that includes perceptions of trust, past experiences, and future or international implications. The case of the trial at Maricopa County (see Chapter 9) illustrated the way in which different notions of the context in question served as a resource for justifying particular evaluations and recommendations.

Beyond just considering the disputes and negotiations about specific devices on the one hand, and the situations in which they have been used on the other, this analysis has sought to highlight the interrelations and mutual definition of technology and its context in accounts about the legitimacy of force. Different contexts or alternative definitions of particular ones evoke different ways of understanding the appropriateness of technology. As well, the understanding given to technology suggests relevant aspects about the context in question. As explained in Chapter 7, because of these dynamics, international legal standards do not provide anything like stable categories and criteria for classifying the legitimacy of particular activities.

Overall, negotiations and disputes about the operation of non-lethals thus have been treated as attempts to establish the proper definition of the context and the technology. Determinations of these matters are always provisional and open to question. In this regard, the major question posed by John Alexander for assessing the merits of non-lethals, 'Compared to what?', is indeed of vital importance, but not just in the sense of comparing the physical effects of non-lethals in laboratory-like settings. Rather, it is necessary to recognize how accounts of the use of force frame the definition of the situation, and just what a particular, weapon should be compared to.

As discussed in Chapters 6 and 9, when things are perceived as having 'gone wrong', the importance of alternative definitions of technology and context becomes quite apparent. When there are deaths in police custody or severe injuries to children in civil disturbances, then reasons for these are sought. Implicit assumptions become explicit articulations. In this, though, there is much scope for the displacement of responsibility (often between competing 'social' and 'technical' explanations) regarding whether particular outcomes were due to the features of technology, the actions of users or those affected, or the unpredictability of the situations in question. For those seeking answers on the behalf of individuals killed, attempts to locate blame with victims, their physiology, or the instability of the events can appear as attempts at obfuscation, designed to shift attention away from anything other than the actions of security personnel or the unacceptability of particular weapons. For those defending such personnel, attempts at making them responsible for situations deemed outside of their control are likewise unacceptable. Ensuing debates about just what went wrong are attempts to simultaneously deliver judgment about a wide range of concerns.[3]

SKEPTICISM AND ANALYSIS

Given the points made above about the negotiation of 'technology' and 'context', the importance of not taking either for granted, and the contention that often follows the use of force, this analysis has adopted a particular orientation to the arguments surveyed. That has been one of 'analytical skepticism', where a question mark has hung over the truth status of statements. This approach has sought to acknowledge and elaborate the constitutive relation between contexts, rules, technologies and practices. Facts were treated as the contingent outcome efforts of persuasion.

Most of the existing writing about these weapons advances an authoritative evaluation of their legitimacy – whether that is positive, negative or mixed. In doing so, definitive facts are marshaled about the actual reasons for casualties, or just how particular weapons have really been used. I suggested something of the contradictory and selective ways certain 'facts' are drawn on to support particular appraisals. There are lots of facts: it is both a fact that the availability of non-lethals has saved thousands of lives in police–public encounters in the United States, and that their introduction has not affected the number of deaths.[4] No doubt a careful and rigorous analysis of the claims and counterclaims in particular cases could suggest grounds for supporting certain evaluations over others. This is a reasonable course for action. Indeed, to some extent, this approach has been undertaken in this book. The skeptical orientation adopted here is not possible without conducting the marshaling of particular accounts of, say, the medical effects of weapons or the motivations of users. In trying to find some way of being skeptical about claims while discussing the controversies on the basis of such claims, this book has taken the pragmatic stance of discussing claims deemed credible, while questioning their status as well. Yet in this, credibility has been judged in terms of the coherence of statements, their applicability and the connections they raise,[5] instead of some incontrovertible grounding.

While for practical purposes the strategy of shoring up a particular version of these weapons is reasonable, I sought to consider if another approach might be worthwhile. The indeterminacies, conditionalities and uncertainties associated with non-lethal weapons make it apparent that asking, say, whether they are really harmless (non-lethal) or harmful (lethal) has its limitations. The answers given are never going to be settled once and for all, because they depend on the manner in which force is used and against whom. Being able to participate in the sort of detailed expert debates that try to separate the harmless from the harmful can require a level of resources and access that are unavailable to most. Formidable demands can be placed on those wishing to assess the most basic statements made about non-lethals. More fundamentally, perhaps, just which account is given – say whether disputes are attributed to 'technical' or 'political' differences – depends on (and helps define) the context in question. Radically opposed versions of non-lethals are offered, even by the same person or organization, depending on the situation.

Examining the basis of claims is an opportunity for acknowledging much wider processes about how technologies get specific qualities attributed to them. When officers are issued with a new weapon and report that it has dramatically reduced assaults against them, but compiled figures indicate no such decrease, then there are pressing questions that can be asked. To ask whether the officers are 'really' mistaken misses that individuals create and sustain divergent accounts that matter in the way force is used, but these are not determined by some intrinsic property of the technology. It should also be recognized that claims about the capabilities afforded by weapons, however much deemed mere hype, offer standards for what counts as proportionate and appropriate responses by helping to define the range of possible actions.

Moreover, the manner in which debates are framed and 'resolved' can be examined in relation to how organizations attempt to secure their legitimacy. What force is acceptable, what justifications are made for it and by whom depends on the distribution of indeterminacies and uncertainties. To the extent that organizations attempt to devise generalized policies about the appropriateness of force that try to be responsive to particular situations, there are important questions about how indeterminacies and uncertainties are managed. While the topic addressed in this book has been non-lethal weaponry, the basic tensions outlined regarding generality, particularity and legitimacy are part of the coping strategies of institutions in dealing with disputes.

The basic skeptical argument is perhaps best summarized by suggesting that what non-lethals do and don't do are products of persuasion based on the interpretation of claims. To the extent that non-lethal weapons are supposed to enhance the acceptability of force, the management of meaning about them will be a matter of some concern. To argue that these technologies are neutral tools, that they help span a massive gap between no force and lethal force, that vigilant media will scrutinize their use, and that they are only being introduced for a limited purpose are attempts to persuade others about the proper basis for understanding technology. The framing given should be questioned.

Take the fact-versus-emotion distinction. Repeatedly in controversies about non-lethal weapons, the debate is depicted as one where facts battle to stem the irrationalities of emotional arguments. Statements that 'most arguments against non-lethal weapons are based on emotion versus facts',[6] or that debates are 'dominated by emotion and preconceived notions of "right" and 'wrong"',[7] are indicative of these characterizations. In such a framing, the facts are almost always taken to be those truths offered by official government agencies. Emotion rests with those seen as presenting some sort of unfounded opposition to non-lethals (relatives of victims, opportunistic lawyers, equally dubious NGOs, etc.). As in so many other areas of life, the distinction between fact and emotion in disputes about non-lethals is drawn with a view to separating legitimate from illegitimate concerns. Facts are on the side of rational and genuine matters, whereas emotion is on the side of irrational and fictitious worries.

Labels of 'fact' and 'emotion', however, often do little justice to the arguments put forward. Characterizing the deployment of plastic bullets (the term itself said to be one with unfortunate connotations) in Northern Ireland as an 'emotive' topic,[8] where the facts must be taken into account, begs the question of just what those facts are and who claims to know them. The failure by some to engage in technical debates about, say, the relative acceptability of injuries sustained from the latest version of a particular weapon in comparison to those from other actions that might have been taken may well derive from doubts about the validity of how official figures are produced. In addition, a number of cases were presented where opposition to the deployment of a certain weapon stemmed from past experiences that were taken to indicate that security agencies were unable or unwilling to use force in the highly proscribed ways asserted. In such an evaluation, the range of issues for consideration are much broader than a focus on the technicalities associated with the latest innovation. In other debates, past practice is taken as a valid basis for future concerns. The opposition to international arms agreements as available option for controlling the development of certain non-lethals is, for some, based on perceptions of the success and failure of past initiatives. What differs between the two usages of past experience is not their logic but how alternative appraisals are made depending on 'who is struggling for credibility, what stakes are at risk, in front of what audiences, at what institutional arena'.[9]

To be sure, though, this skeptical approach has not been neutral in terms of its implications for the different views about non-lethals. In unpacking the 'facts of the matter', most of the attention has been given to understanding the contingencies underlying optimistic assessments, as these are most prevalent and supported. For those most committed to marketing the benefits of non-lethals, to discuss their worth at all as 'debatable' would itself be recognized as a major and unwarranted commitment. Nor has this approach been neutral in its treatment of its implications for different organizations. To the extent that police and military organizations are the most prominent and treated as the most credible public commentators about the appropriateness of force, to question the basis of claims made is necessary to question this status.

Earlier chapters highlighted the limitations of analysis, the contingencies of accounts of non-lethals and the work necessary for the presentation of certain facts. These points are not just relevant to assertions of others scrutinized in this book, but also apply to my claims. In discussing the legitimacy of non-lethals, I – just like other commentators – have done so by drawing on prior interpretations of events made by others. This condition raises significant questions about the need to consider the subsequent interpretations offered.

The attempt has been made to display the effort undertaken to substantiate the claims and generalizations made by others and myself and to comment on the contingencies therein. The tension that runs throughout has been how to struggle against the closure of debates in order to take a skeptical stance toward the claims made, while necessarily implying a certain reading of events through the accounts given. In considering disputes as such, the meaning of social action has not been treated 'as a unitary characteristic of acts which can be observed as they occur, but as a diverse potentiality of acts that can be realized in different ways through participants' production of interpretations in different social contexts'.[10] It is difficult to see how an analysis that took seriously the importance of alternative interpretations and contentions about legitimacy could sensibly proceed without recognizing the conditions of its interpretations. While this may not pave the way for the sort of loose and fast claims that sell gadgetry, it points toward a basis for understanding the commitments of analysis and suggests various lines of caution. Where questions about the interpretation of events have been displaced to the reader, the attempt has been made to highlight the importance in disputes of what questions are asked and how those are answered.

RECOMMENDATIONS

In countries wishing to uphold democratic principles, the use of force is problematic. While, undoubtedly, agencies of the state have just cause to resort to it at some time, any such action is likely to raise questions about its appropriateness. To say that the use of force must always be proportional to lawful objectives or that restraint must be exercised is not to provide a solution to the questions about what ought to be done, it is rather to give a particular articulation of key areas of concern. As weapons ostensibly designed and used to minimize deaths and injuries (and therefore demonstrate proportionality and restraint), non-lethals do not escape from tensions associated with force. If the safest non-lethal weapon is a marshmallow dropped by parachute, then it is also one of the least effective. To compel certain forms of behavior or gain control of situations through force is never likely to be far from controversy. When non-lethals are expected not just to maintain the legitimacy of force but also to enhance it, then the goals set are high indeed. Questions are likely to be voiced, and in democratic countries should be voiced, about whether other alternative-force options or other actions could have been pursued.

The analysis given in this book has sought to further an appreciation of disputes about these risky technologies. Disputes often center on the meaning of classifications and characterizations and which of these ought to be offered. To acknowledge the conditions noted in the last section is not to fall into a state of paralysis regarding more practical issues, but to approach these in a certain way. Making a case for the dilemmas associated with evaluating claims and the potential for controversy suggests that while recommendations are advanced, their limitations need to be considered as well. In the remainder of this section I want to offer recommendations about approaching the recourse to non-lethals. As a diverse range of considerations must be brought into play while examining the deployment of non-lethal weapons, so too these recommendations cover a varied terrain.

Take the Assessment of the Effects of Weapons Seriously (But Not Too Seriously)

Previous chapters suggest that many of the controls in place to assess the safety of weapons and justify the deployment of non-lethal weapons have been wanting. It is impossible to accurately characterize the varied regulatory regimes in place internationally in a simple statement. Yet, a number of cases in different countries have been examined where plausible reasons exist for doubting whether regulation agencies met the standards of rigor these organizations set out for themselves. At its most extreme, pronouncements about the latest weaponry innovation have claimed that it offers solutions to the problems experienced hitherto about the use of force. Statements to this effect are often in contradiction with other widely held beliefs, the arguments put forward are internally inconsistent, and the facts given are subject to significant but unstated qualification. Examples have been given of highly optimistic claims that are taken as given initially only to become subject to major doubt later. Less dramatically, uncertainty is rife about the possible effects of future weapons and the effects of past deployments. Yet, in many cases, the lack of definitive evidence is taken as proof of safety.

Taking past experience in this book as a valid – but contested – input into evaluations, one general lesson is that unsubstantiated claims about

259

the merits or drawbacks of particular weapons made by manufacturers, governments, or others should not be relied on as they have in the past. As further suggested, though, even when particular statements of safety are supported with some evidence, a skeptical orientation is prudent. The detailed examination of various 'facts' about particular weapons gave significant reasons for doubting their robustness. The validity, conditionality and the scope of claims are matters for detailed consideration. Suggestions that weapons have been rigorously tested desire close scrutiny in regard to just what has been tested and in what manner. A pronouncement by a prestigious government advisory committee about the safety of a weapon, for instance, is not the end of safety disputes, but is yet another step in how the factual status of a certain claim is built up.

Given these conclusions, there is considerable scope for enhancing the level of testing and regulation associated with non-lethal weapons. The call has been made from many quarters – including defense companies, NGOs and medical researchers – that non-lethal weapons should be assessed in a manner akin to the assessment of pharmaceutical drugs. Especially in relation to weapons being proposed today that seek to inflict injury in novel ways, the general thrust of this suggestion would seem to have much merit in reducing some of the unknowns and uncertainties about non-lethals.

To contend that weapons should be treated like drugs with respect to their regulation, though, is more to signify the desirability of enhanced control than to point to universal procedures that could simply be put in place. Just what, specifically, this suggestion would mean requires some consideration. In general, the approval process for licenced drugs consists of four phases: Phase I involves animal and lab tests for toxicity and safety; Phase II consists of assessments of safety and efficacy in a small sample target population; Phase III evaluates the comparative efficacy of a drug against existing treatments or placebos in a much larger sample; and 'Phase IV' consists of post-approval monitoring of adverse reactions to a drug as it is used in practice. While many of the efficacy goals of therapeutic testing do not make sense in relation to weapons, what the general procedures in place for drugs suggest is the importance of the pre-approval testing of known risks and the post-approval monitoring of effects as part of licencing arrangements. Such overall prescriptions seem quite sensible for non-lethals. Regulatory controls could be tied into licencing procedures that sought to ensure that the claims were substantiated.

As suggested in Chapter 5, though, the adequacy of existing drug-approval provisions is itself a topic of some public debate. Much of the concern centers on how to ensure the independence of testing. Along these lines, John Abraham has offered various recommendations for improving the drug regulation procedures:

- Control of drug testing and trials should not be only within the control of manufacturing firms; regulatory agencies should play an active part in the testing required;
- Those making licensing decisions should not be reliant on industrial expertise;
- Regulation should take place outside those institutions responsible for promotion and sponsorship of drugs;
- Public scrutiny through Freedom of Information and other means should be available during regulatory reviews as well as after. This should include detailed transcripts of the decision proceedings of regulators;
- Uncertainty about risks should not be used as an excuse for inaction or approval of products of uncertain worth.[11]

These suggestions provide useful points for thinking about what procedures could to be undertaken. Much of the current testing for non-lethal weapons in the United States and Britain, for instance, is being undertaken by those government departments most actively supporting their development.

Underlying Abraham's recommendations is the importance of shifting current distributions of the benefit of doubt in approval decisions. This has been a key area of concern in this examination as well. Generally, the current overall lax state of controls means the responsibility falls on those seeking to prove particular weapons have unacceptable risks rather than on those promoting the technologies to provide a substantiation of safety. Evidence of injury compiled after deployments begin are likely to be inadequate and equivocal. In situations where problems are only likely to emerge after many years, as with chemical irritants, the approach of waiting for injury has serious drawbacks. In this way, calls to consider the distribution of proof share much with 'precautionary' approaches to hazards that stress the importance of preventing unacceptable risks and acting to anticipate problems before they arise.[12]

The transparency of decision-making processes is also a fundamental aspect of any approval procedures. Access of information has been a central issue in the controversies studied here.[13] The ability to offer credible claims is often predicated on access to information. Without transparency in decisions it is too easy for those in control of the 'facts' to define issues away. Various examples have been given where important material has been deemed secret, made accessible to only a privileged few, or has been highly 'edited' prior to its becoming public – acts that, together, can be referred to as the suppression of information.[14]

Consider the points above in relation to the CS sprays in Britain. As part of a review of less-lethal weapons for the police in Northern Ireland and mainland Britain, ongoing at the time of writing, a Steering Group led by the Northern Ireland Office has been set up to review current options. Phase II of that review advised the extending of the deployment of CS sprays to routine policing in Northern Ireland. To deflect potential safety concerns discussed in Chapter 8, the Steering Group 'acknowledged that there have been one or two reports suggesting that MIBK is not the ideal solvent, and that it might carry risks that another solvent might not'. Presumably the 'one or two reports' mentioned refers to the two Porton Down studies, detailed in Chapter 8, which argued that MIBK had an unacceptable carcinogenic and mutagenic risks. Taken as isolated studies, their importance might be limited. If, for instance, there had been a number of other relevant studies conducted as part of the approval process, then the weight of the studies could be downplayed. According to answers given in Parliamentary correspondence I have obtained, though, these were two of the three main health evaluations undertaken about the sprays for some time after their approval. As suggested, the third being of somewhat questionable relevance for assessing CS *sprays*. Because the specifics of the approval process for the spray had not been made public, though, it was impossible for most to situate the Porton Down reports and understand their importance with the approval process.

Furthermore, in the case of CS sprays, the lack of public specification about how the Himsworth Committee have been interpreted – whether the recommendations to test akin to a drug refers to 1971 standards or current state-of-the-art – would suggest that the meaning of criteria, as well as the evidential basis of decisions, should be elaborated in a manner accessible to concerned parties. This includes individuals such as police chiefs, who will be liable for deployment decisions.

The sort of selective portrayal of information, as exemplified in the Northern Ireland Steering Group review above, does little to enhance confidence in decisions taken, at least for those able to situate the claims offered. Admittedly, in many cases few officers, citizens or others are likely to be able to undertake this. Yet, when long-time disputes about the effects of a certain kinetic weapon have centered on past injuries and deaths to innocent bystanders (such as in Northern Ireland), failure to publish commissioned assessments of the latest round cannot be helpful in boosting public confidence. When the scope for interpretations about the merits of force is curtailed in such a manner, then critical commentary is warranted.

The opportunities afforded for transparency of safety assessments and other review processes differ markedly from country to country. Chapter 9 drew on a number of documents provided by the Freedom of Information Act in the United States, whereas possibilities for requesting such information through the UK Open Government scheme are much more limited. Following Abraham's recommendation, scrutiny of review procedures through Freedom of Information and other means should be available both during and after the period when decisions are taken. Prior to deployment decisions, the assessment basis for decision should be published in the open scientific literature. Information about who makes decisions and any conflicts of interest should be available.

If those in charge of assessment procedures do increase their transparency, one of the likely results is that critical attention will be cast on those organizations most open. It is the fleeting glimpses provided by a leaked report, or a lawsuit, an act caught on tape that provide much of the basis for debate. It is possible to think about the evidential basis and contingencies of the Dutch review of pepper sprays because the Netherlands, in contrast to many others, publicly detailed its appraisal. Given disputes about the credibility of evidence and the implications of limited data and understanding about many medical effects, 'merely' describing what goes on (and what does not go on) in assessments is sometimes enough in itself to cast doubt on their rigor. For those seeking to improve decisions about the use of force, while perhaps unfortunate, the skewing of attention should not be a barrier to reform. As one senior tactical support police officer who commented on a draft of this book said: 'Discussion about weapons is GOOD. I love it, because it makes democracy work. I want darn difficult questions asked because they make me rethink unless I can give a good answer.' If only this attitude were more widely shared.

The desirability for thorough and independent assessments of non-lethal weapons seems obvious in alleviating major hazards. To be sure, lessons from past experience would suggest that such work has not always taken place. Yet, the importance of testing effects is commonly recognized in international discussions today. A wide range of organizations – police forces, human-rights groups, security analysts, etc. – cast many of their central concerns about non-lethals in terms of the need for rigorous assessment. Appeals to 'science' are seen as credible, never mind that science is multiply conceived and it is drawn on to support opposing conclusions. With funding streams of tens of millions of dollars becoming available in the United States and other countries, defense establishments, universities and police agencies are lining up to be commissioned to undertake such work. Just what counts as 'thorough' and 'independent', what effort should be dedicated to such activities, and how evaluations ought to enter into approval and licencing decisions are likely to be topics where consensus breaks down. This book has stressed the importance of questioning what is expected from such testing. The facts do not come forth from appraisals and point to the unambiguous way ahead. To suggest that particular tests could provide a firm grounding for legitimacy misconstrues the ability of analysis to resolve disputes about force.

The limits of analysis are numerous and varied. To start with, experts disagree. Despite years of research into current 'off-the-shelf' technologies, the acceptability of various incapacitants or kinetic weapons are matters where authoritative analysts and government agencies have come to opposing evaluations. Those seeking to adjudicate disputes by calling on 'the experts' have important questions to ask about which expert to ask and how this might affect the conclusions reached. Additionally, instead of resolving disputes, research can promote them by bringing up further uncertainties, as one claim provokes several lines of possible concern. As suggested earlier, the indeterminacies and uncertainties associated with effects mean there is often ample scope for making alterative assessments. Moreover, knowledge is often late in coming. The examination of chemical irritants illustrated that problems are only identified after many years and proving causation is quite difficult. Various 'definitive' and expensive investigations have been examined here, and significant questions were expressed about their value. When tests are done and inconvenient conclusions are reached, analysis has been, and can be, ignored. The meaning of standards of assessment that are supposed to define what counts as good testing is continually up for negotiation. In Britain, for instance, in every case of the introduction of a new chemical agent into police and paramilitary units since the 1970s (CS sprays, PAVA sprays and CR gas), there are plausible grounds for maintaining that the procedures supposedly governing procedures have not been followed. What this should be taken to imply for the believability of future claims is likewise not a matter settled by conducting certain expert studies.

Testing and evaluation will also not answer questions about the weighing of risks and about what level of force-response is appropriate for certain threats. Any use of force in policing, peacekeeping and other such activities requires balancing the safety of security personnel against concerns to act proportionally. If the safety of security personnel was all that mattered, then any force-response thought effective could be justified. If the safety of those targeted was the only issue, then perhaps no force at all could be. In between these two extremes, though, there are thorny questions about how the lives of bystanders, security personnel, suspects, combatants and others should be balanced in thinking about what constitutes appropriate action. No amount of testing can resolve such questions.

Even assuming that a particular device entails little or no risk of long-term physical injury does not necessarily make it acceptable. Medical and safety evaluations also give little voice to the pain inflicted by non-lethals. Such subjective experiences are generally rooted out of the language of official evaluation reports. When the police employ a chemical spray that allegedly stops misbehavior in seconds but induces intense pain for fifteen to thirty minutes and no decontaminant is made available, then questions can be asked about the ends served. Much of the current research into malodorants and other next-generation weapons appears geared toward maximizing pain.

In this book, the importance of the interrelation of claims about the effects of a weapon to its context has been highlighted. While laboratory tests or controlled field experiments might provide some insight, the information gained is likely to be of questionable worth in commenting on controversies experienced in practical applications where such artificial conditions do not hold. Assessments made in relation to a specific context can have their value come under question if the situations of use diversify. But the interrelation of technology and context is not simply a one-way relationship where context shapes what effects are anticipated and acceptable. Determinations about the acceptability of weapons rely on an understanding of the context in which they should be understood, but at the same time the important aspects of the context for evaluation derive from a sense of how the technology would operate in that setting. This iteration is key in the unfolding of debates about the merits of weapons.

Approach Non-Lethals as Elements of Socio-Technical Systems

These points about technology and context signal the importance of attending to varied issues in thinking about the legitimacy of the use of force with non-lethal weapons. To frame debates in terms of the characteristics of technologies alone – for instance, by seeking to establish probable effects – is to miss how they function in practice. Previous chapters illustrated how uncertainties and risks about the deployment of these weapons expanded, contracted, became redefined and sometimes defined away in specific settings. The development and deployment of non-lethals is a socio-technical exercise that should not be reduced to a study of technology in isolation.

Consider these points in relation to a specific policy-evaluation process. Much of the ongoing review of 'less-lethal' options by the UK

Steering Group in its Phase I & II reviews has been devoted to the predictive assessment of effects. Yet, such evaluations are just one part of ensuring the appropriate deployment of less-lethal weapons. As with any technology, there is the potential for unforeseen risks that no amount of precaution can completely avoid. Further, though, medical and technical assessments, no matter how rigorous, cannot in and of themselves guarantee the acceptability of weapons in practice. At the very least, mechanisms need to be in place to monitor deployments and tactics so as to revise actions on the basis of experiences. Numerous examples were given in previous chapters where such mechanisms were absent. Attention of the Steering Group has been on assessing technical innovations rather than on considering wider institutional topics that speak to more diverse social and technical issues.

A few examples will serve to illustrate possible options. Physicians in southern California examining hospital casualties from police use of a beanbag (see the third section of Chapter 5, pp. 97–8) contended that a pattern of underestimating injury existed and advised on the need for more robust assessment systems than had been envisioned. In relation to CS sprays, the UK Police Complaints Authority recommended that all police forces should record injuries to the public in use-of-force reports. In further relation to the sprays, some police forces routinely monitor levels of the irritant in officers' sprays in order to estimate the amount used in encounters when allegations of excessive force are made. With regard to the baton round, the Northern Ireland Patten Report into policing recommended that, wherever possible, a camera recording be made of incidents involving plastic-baton rounds. Yet, such organizational reforms typically do not play much of a role in official evaluations of safety.

The analysis here suggests that it would be advisable to consider a range of reforms in the way the use-of-force reports, injuries, and compensation claims are monitored, as well as how such information is made available. Such feedback measures can help increase the flexibility of policies. The importance of maintaining flexibility is a key theme in the social science literature on technology assessment. For instance, Collingridge contends that a fundamental dilemma exists in the control of technology: when that is relatively easy in the early stages of development, it cannot always be justified because negative effects are not readily apparent. Yet when the need for control is apparent, it is often expensive and more troublesome because the technology has become embedded within organizational practices.[15] While this analysis has sought to eschew definitive claims about when control is really possible in favor of noting how such a space is negotiated throughout the development and deployment of technology, the emphasis on encouraging flexibility is still prudent.

266

In light of previous arguments, it would be unwise to suppose that measures such as those mentioned above would yield authoritative information. The triumph of facts in resolving disputes about force is a long way off and it will be so no matter how much data is generated. Take, for instance, attempts to increase the accountability of electroshock weapons. One of the most contentious issues about such weaponry is their lack of easily identifiable marks and the related scope for abuse. To enhance the likelihood of arrest should the weapons fall into criminal hands, one TASER company has developed a system for dispersing small identification tags from its weapons when they are fired.[16] Each device has an individual identification number on the tags, and information about the personal details of purchasers are supposed to be recorded through manufacturer-retailer links. In addition, the latest Advanced TASER contains an onboard computer that records the time and date of every firing so that police agencies can monitor usage patterns. Taken at face value, such initiatives are laudable. Yet, in practice, there are pressing questions about how such information would be made use of, who would have access to it, and how conclusions about controversies would be made.

The points made in this section about seeing non-lethals as part of wider systems is not to suggest that '[b]y isolating technology as *the problem*, we miss the point that people are really at fault'.[17] As argued, to pose the question in terms of whether it is the technology or humans that are at fault is a particular, contingent and, in many ways, unhelpful way of talking about the issues at stake. Such a framing obscures the way weapons get taken up as part of wider assemblages or systems and how fault is negotiated in practice.

There is the obvious danger in conceiving of the use of non-lethals in terms of wide-ranging socio-technical systems that notions of agency or responsibility might become completely lost. As this analysis has suggested, there are no easy answers to what constitutes the bounds of relevant considerations that ought to be brought to bear in examining technology – whether this relates to factors about particular instances or the importance of past experience. However, the focus on the issues mentioned in this section would help to pose a variety of issues not typically given much attention in the policy discussions surveyed in this book.

The Processes of Attributing Responsibility Need Careful Attention

Let us turn, then, to a discussion related to the extent to which 'people are really at fault'. In the abstract, attributing responsibility in the use of force should be relatively straightforward. In relation to concrete cases, however, any number of faults might be offered to explain why actions did not develop as expected and such accounts lead to alternative determinations of who (if anyone) or what (if anything) is to blame. While it is not impossible to find particular instances of force that are widely praised as exemplary or condemned as abusive, often their appropriateness is disputed. If everyone shared a sense of the threat posed in certain situations and of what effects were probable from particular actions, then it would be easy to resolve disputes by reference to good or bad intent. But such an understanding is far from guaranteed.

The analysis here has viewed that the use of force with non-lethals is highly dilemmatic. To start with, it is obviously inappropriate for security personnel to take any action they see fit, with no regard to standards of conduct or scrutiny by others. Even in war, rules are said to apply, however fragile. Non-lethal weapons are supposed to be utilized in highly proscribed ways. Yet, no matter how well internalized, rules cannot dictate appropriate action in every circumstance. General rules must be applied in specific situations which are never identical. Because rules about the use of force can never fully exhaust the range of circumstances they are trying to comment on, they are insufficient to completely direct action. Rules about the use of force are qualified by various clauses, 'ifs' and 'buts', which mean that questions about whether they are, could or should be adhered to are often matters for disagreement. Complex questions about what force should be used and when cannot be resolved by rules written in police headquarters or diplomatic conference halls in Brussels. Just what rules should be adhered to in the first place depends on how situations are defined. These points suggest that when police policy-makers give out guidance that 'merely' suggest how rank-and-file officers should act, when government agencies 'only' advise about the safety of a certain device, or when researchers come to findings that are not recommendations but instead 'guides' to others, then questions should be asked about how such recommendations are taken forward.

Much of the existing public discussion about the use of non-lethal weapons centers on the adherence (or lack of it) to rules. In official public-relation pronouncements about how non-lethals will and do function, it is often claimed that rules are to be strictly adhered to. Yet, this public rhetoric is often difficult to reconcile with the confused and confusing world of practice. As illustrated previously, in relation to specific controversies about the appropriateness of action, though, such statements quickly lose their salience. Personnel are said to have to act in a split second, where there is no time to follow the niceties of bureaucratic regulations. In controversies, the lack of rule-adherence might be seen as stemming from the general volatility of situations rather than from malicious intent. In situations where discretion is simultaneously deemed central and denied and where it operates as a resource for cutting off disputes and for infinitely displacing blame, then, contention will not be far off.

In order to provide greater clarity, some suggest that legally binding rules of engagement should be placed on the employment of non-lethals.[18] Such calls are understandable and not without some merit. Allegations of acts of gross excessive force might well receive an easier airing when rules have a firmer legal footing. There is much that could be done, for instance, in the way of setting out expectations for after-care. This analysis, though, has suggested a number of reasons for doubting how far it would be possible to devise and enforce detailed rules.

If discretion of individuals who use force cannot be completely elim-inated, then the dangers associated with it can be reduced or structured in such a way as to minimize possible concerns.[19] One way of doing this is by forgoing those weapons that require adherence to strict rules. The deployment of incapacitant sprays that should not be used at distances under a meter or in enclosed spaces, even when these are the types of situation in which past usage suggests they are needed; kinetic weapons that must not be fired under a certain distance even when users admit to having difficulty estimating distance; and electroshock devices which should only be applied for a limited duration but where their effective-ness cannot easily be assessed, are of questionable merit. Of course, given what has been said in this book about specifying the characteris-tics of technology, the process of evaluating weapons against particular standards is not likely to be straightforward. Rather, the suitability of a given weapon and to what extent it might be problematic must arise out of an iterative process that moves back and forth between an under-standing of technology and its context.

Beyond these issues of the appropriate technology, this book has stressed the importance of how determinations of responsibility are structured within and across organizations. As suggested in Chapter 8 (see pp. 192–3), in practice the division of expertise can be split in such a manner that finding anyone willing to take responsibility for decisions or actions is a major achievement. There is much 'gray space' about force and rules, but this should be seen as a product of the way organi-zations handle indeterminacies and uncertainties.

One line of consideration, then, is to ask whether there are alterna-tive ways of distributing responsibility, so that discussions do not degrade into the types of oppositions (use vs abuse; accountability vs culture of blame) that often characterized the debates surveyed previ-ously. For instance, elsewhere I have discussed a system devised by one police force in Britain where trainers systematically collect information on force incidents and incorporate the lessons of such reviews into training provisions.[20] In doing so, the instructors are able to provide more than mere guidance, but instead offer accounts of what types of force and equipment ought to be used in what circumstances, how potential conflict situations should be approached, and how the application of force ought to escalate. These suggestions are informed by various procedures for monitoring public–police incidents and setting up procedures for learning from experience. Complaints against an officer can be situated in relation to his or her history of the use of force, and officers can be referred for more training when deemed appropriate. Ideally, the information collected can be used both to protect police and to take action against inappropriate behavior. The crucial point is that the system offers a different approach for how and where responsibility is distributed. It is not the case that, in this, the basic disagreements about the acceptability of force disappear. Street-level officers are still legally accountable for their actions and have a broad latitude of discretion, but policing takes place within a set of procedures designed to help senior and junior officers learn about force options, as well as punish acts deemed inappropriate. Rank-and-file officers must be able to justify deviations from those procedures deemed to be most appropriate, rather than merely making ad hoc determinations of the best response in specific cases. For their part, senior officers should know what are the current and emerging use-of-force problem areas and suggest measures most likely to be effective in dealing with situations. Whatever the practical limits of any such highly rationalistic system, the tendency for evaluations of force to degrade into simple stories of street-level officers' use and abuse of a technology is generally less likely because the figure of responsibility points to senior officers as well.

Recognize the Conditions of Knowledge Production

This book has examined the basis and adequacy of competing accounts of particular events in a fair amount of detail. In asking what recommendations might follow, then, it is important not only to consider how appraisals about safety or other issues are made, but how discussions about non-lethal weapons proceed. In the controversy that often surrounds the use of force, the manner in which events are portrayed – who uses what language, how, when, and for what audience – is a topic of some importance.

Much of the commentary in this chapter has been directed at government agencies and manufacturers. Just how allegations of excessive force are made is significant as well. Chapter 9 considered something of the contingencies and limitations associated with accounts given by those providing outside scrutiny to security agencies.[21] Wilson, for instance, argued that human-rights organizations often adopt a legalistic basis for authority that strives for objective facts when the possibility for such knowledge is doubtful. In an effort to persuade governments, unrealistic divisions are sought between facts and values as well as objectivity and subjectivity, where human-rights organizations firmly side with the first of each pair.

Similar themes have been developed in this analysis of non-lethals. At best, the pragmatic search for the 'facts of the matter' to guide evaluations is tension-ridden. The analysis of pepper sprays and stun devices in Maricopa County jails illustrated something of the elusiveness of steering facts and an objective grounding for evaluation. The effects of many non-lethal weapons are conditional, as well as contested and contestable. The somewhat unavoidable response to the question 'Is X weapon safe?' is 'Well, it depends'. Determining just what it depends on and whether those conditions have been met is quite demanding. Few (if any) organizations have the access, time or resources necessary to gather the sort of basic information necessary to comment on the range of possible factors at work. As suggested here, even when some are commissioned to do so, the results can be quite questionable. A reading of the intent of actors is similarly difficult to substantiate in an authoritative fashion. Attributions of 'good' or 'bad' intent of security personnel might help clarify discussions about force by providing a clear decoder of why certain actions took place, but it is questionable how adequate such attributes are at capturing the complexity and diversity often at hand. As suggested at various places in this book, determinations of effects and intent can, in many ways, be understood as products of particular lines of argument and ways of handling uncertainty. Moreover, the conflicting accounts given in the Maricopa County case indicate how the claims offered about what happened were related to the position of those commenting vis-à-vis the events in question. The definition of 'the problem' is bound with the knowledge possessed. For human-rights groups seeking definitive facts while being cut off from the sort of access that grounds others' credible claims to 'the facts', their ability to offer persuasive counter-claims can be quite limited.

As such, this book has shared Wilson's desire to examine the 'powerful representational claims articulated within human rights discourses, so as to drag them down from the rarefied epistemological and moral high ground, and include them in more sociological debates about interpretation, understanding and explanation of empirical evidence and the limits of its representations'.[22] He suggested that human-rights organizations and others could adopt a wider range of representational styles by, for instance, further incorporating subjective accounts within reports, or placing individuals' experiences within their situational context. These remarks are not meant to downplay the importance of denouncing abhorrent acts. Rather, it is because so much is at stake that the types of story told deserve close attention.

This analysis of the use of force with non-lethals does not just suggest the need for a greater contextualization of accounts, even if one acknowledges that there are likely to be multiple and competing conceptions of the proper 'context'. Rather, I have sought to highlight the ways in which technology and its context are mutually defined. In practical decisions about the legitimacy of the use of force, weapons are interpreted in relation to their context and, at the same time, the context is understood from descriptions of weapons.

As part of questioning the legitimacy of force in those cases where much uncertainty exists, then, one way forward in accounts is to make explicit this interconnection and the possibility for multiple interpretations of just what is going on. Accounts can evoke an acknowledgement of the possibility for alternative characterizations of the proper context for evaluation, the technology, and how determinations of one affect the other. In this, offering an evaluation of the acceptability or unacceptability of non-lethals can make more explicit the basis and reasoning for assessments of the appropriateness of force. While it seems unlikely that such actions would resolve competing accounts, they may provide a basis for acknowledging and communicating further sources of concern than are voiced in a call for the facts. This approach can provide a basis for asking further questions regarding official stories of what happened.

Casting attention in this manner recognizes the likely endemic character of disputes. As such, the possibilities of substantiating the benefits of the introduction of particular weapons becomes highly contingent. Reaching a confident conclusion regarding the relatively benign status of force with non-lethals may require that a number of things function as advertised: the robustness of the initial scientific and medical evaluations, the rigorousness of the training procedures, the adherence to proscriptions, the perceptions of users regarding the utility and dangers of non-lethals, the functioning of surveillance systems for tracking situations of use and likely implications of technology, the thoroughness of the monitoring procedures for complaints procedures, the adequacy of personnel provisions, the expectations surrounding the benefits of non-lethals, etc. Just what might be required, though, by way of procedures or levels of testing depends on an understanding of what non-lethals can and cannot do and of just how dangerous they are. In the difficulty of

272

pinning down technology and context, there are pressing questions about what precautions are taken with the use of force and whether lessons have been learned from past experiences. These points about the indeterminacy of events are not to suggest that all accounts of what happened are equally valid. Robust knowledge of the merits of non-lethals should be able to acknowledge the possibilities for multiple definitions of technologies and context, but offer a persuasive reading of the two together.

Perhaps more importantly for the line of argument here, recognition of the interdependence of context and technology underscores the importance of displaying how certain determinations of praise or blame are made. In this, what is being said and who says it are interdependent. An acknowledgement and incorporation of these issues into assessments made provides a basis for pointing out the contingencies of assessments and their dependence on conditions of access and other matters that might be worthy of critical attention. Questions about the legitimacy of force can be treated not simply as a matter of specifying effects or merely noting the scope for conflicting assessments, but instead as one of building an approach sensitive to the conditions under which interpretations are made and to how the legitimacy of claims are secured.

Constantly Ask What Else Might Be Done

If one takes the use of force as a necessary evil, or at least as an act that requires the balance of various tensions, then continuing questions can and should be asked about its appropriateness, as well as how this is determined. In terms of ensuring both public support and responsible conduct, it is unlikely that any amount of research, testing, training or monitoring can substitute for the public accountability of security organizations. In Chapter 6, for instance (see p. 141), the lack of sanctions brought against personnel in Northern Ireland and Israel for perceived gross transgression of guidelines has been a source of much discontent. It would seem unwise to attribute such continuing disquiet about non-lethals to a failure by the proper authorities to sufficiently communicate their real merits. Wider questions about the contrast in power between accuser and accused, say in relation to the inequality of means for security agencies and victims to substantiate their interpretation of events, appear much more relevant.[23]

Given the inability of expert analysis to resolve controversies about non-lethals and force, there are vital questions about how to improve decisions. As political scientists Lindblom and Woodhouse argue, the intelligence of decision-making is largely achieved by the interaction of different groups in a manner that brings to the fore competing contentions.[24] Enhancing decisions about public policy requires learning from experience, proceeding with policy in a flexible manner, and striving for a greater equality in the ability to raise concerns.

A number of suggestions follow from the spirit of these observations. First is the importance of proceeding cautiously. This might require, for instance, protecting against hazards by initially placing limits on the development and use, as well as protecting against severe risks. Initial tests, such as those in artificial and controlled environments, might be helpful. A key issue is whether these are conducted to provide a source for future learning or a means for closing down discussion. Additionally, the inability of experts to resolve disputes about causation should provide some caution against pursuing ever-new force options. Also, it has been suggested in this book that the area of weapons proliferation remains one where easily identifiable further steps can be taken to mitigate the contribution of non-lethals to human-rights violations.[25] It would be a tragedy indeed if the next generation of non-lethals were as freely available as are existing forms. Similar cautionary lines might be explored. Much of the latest non-lethal weapon research proposes to target the body in specific and novel ways. The possible long-term consequences for the future of conflict are immense. While claims about imminent forthcoming technology must be treated carefully, it is also wise to ask what preventive action might be taken to reduce the threats identified. Here, the strategy of placing restrictions on certain initiatives pending further developments might offer significant merit. Initiatives such as the SIrUS project, while not providing criteria that resolve categorization disputes about technology and context, are valuable for establishing terms for debating the acceptability of particular initiatives.

Second, calling for enhanced assessments or policies is not to imply that decisions should be left with scientists, security personnel or legislators. No group in democratic countries should be the sole best judge of standards and the acceptability of their practice. This analysis would suggest that improving the competition of ideas and the interaction in political debates are key with respect to decisions about non-lethals. The ability of corporate and government agencies to make highly optimistic claims that are impervious to challenge should be countered. Improving decisions made about non-lethals requires promoting debate and the possibility that optimistic and pessimistic claims will be open to scrutiny. Take the issue of weapons proliferation again. Recent questionable arms transfers in Europe and North America, brought to public attention by journalists, NGOs and others, have embarrassed some governments into putting into place basic safeguards and monitoring procedures for arms transfers. Such criticisms play a vital role in minimizing the worst violations of international law.

Certainly, one of the most significant questions where alternative ideas should be brought to bear is whether force is necessary at all in dealing with potential conflict situations. Various examples have been given in this book where the resort to force is justified on the basis of its 'non-lethality'. That supposedly 'non-lethal' options are available should not become a justification for ruling out other options. There is scope for novel thinking about alternative approaches. For decades, South Korean police were pitted in battles with students, workers and protestors. Massive amounts of tear gas were used as a means of 'resolving' such conflicts. In recent years, though, police forces have turned to unarmed female desk officers, operating with non-aggressive tactics, for the frontline policing of demonstrations.[26] The pitched violence and injury that marked protests in the past is said to have almost completely stopped. Whereas in 1997 there were 220,000 tear-gas canisters employed, in 1999 there were none. As in any conflict response, though, there have been certain casualties. The Seoul-based company producing tear gas has suffered a major reduction in sales and male former riot-control officers have been transferred to other jobs. Just why conflict seems to have been reduced so dramatically and what lesson there might be for other countries are topics worthy of consideration. It will probably be of little surprise, though, that this Korean example has not been widely, if at all, discussed among those seeking to promote the latest gadgetry. This sort of initiative, that seeks to reduce the resort to force, is in line with the suggestion by the Chief Constable of the UK Hertfordshire police force, Paul Acres, that 'talking is the most effective way of dealing with situations'.[27] There are many other possible avenues for novel means of policing and maintaing social order.[28]

In the end, there are no easy answers to controversies about the legitimacy of force with non-lethals, or even about how best to approach this issue. This book has underscored the importance of constantly questioning how politics and technology are understood. A consideration of non-lethal weapons illustrates the hopes and fears associated with technology, and how it functions as a site for struggles over meaning and morality. It is because the relation between technology and politics is so debatable that discussion needs to continue.

NOTES

1. Office of the Director of Defense Research and Engineering, *Report of the Task Group on Biological and Chemical Weapons Development of the Defense Science Board*, DSB 225/3, 18 February 1959, two vols. Quoted from Robinson, J., 'Disabling Chemical Weapons', Presentation to PUGWASH Study Group on Implementation of the CBW Conventions, 27–29 May (Den Haag: PUGWASH, 1994).

2. See for instance Sautenet, V., 'Legal Issues Concerning Military Use Of Non-Lethal Weapons', *Murdoch University Electronic Journal of Law*, 7, 2 (2000).
3. Bhatt, R., 'The Extent of Problems and the Families' Perspective', Presentation at Safer Restraint Conference, London, 17 April 2002.
4. See, respectively, Mangold, T., *Panorama – Lethal Force* and Bailey, W., 'Less-than-Lethal Weapons', *Crime and Deliquency*, 42, 4 (1996), 535–53.
5. For a discussion of a non-objectivist basis for evaluating arguments see Herrnstein Smith, B., *Belief and Resistance* (Cambridge, MA: Harvard University Press, 1997).
6. Alexander, J., 'The Future of Non-Lethal Weapons', *Medicine, Conflict and Survival*, July–September, 17, 3 (2001), 190.
7. Smith, R., (1997) 'Reducing Violence', proceeding from *Security Systems and Nonlethal Technologies for Law Enforcement*, 19–21 November 1996, Boston, MA (Bellingham, WA: SPIE), 27.
8. Acres, P., 'Conflict Management Portfolio', Presentation at Safer Restraint Conference, London, 17 April 2002.
9. Gieryn, T., *Cultural Boundaries of Science* (Chicago: University of Chicago Press, 1999), ix.
10. See Gilbert, N. and Mulkay, M., *Opening Pandora's Box* (Cambridge: Cambridge University Press, 1984).
11. See Abraham, J., *Science, Politics and the Pharmaceutical Industry* (London: UCL, 1995).
12. For a discussion of the problematics associated with this principle see Jordan, A. and O'Riordan, T., 'The Precautionary Principle in UK Environmental Law and Policy', in Gray, T., *UK Environmental Policy in the 1990s* (Basingstoke: Macmillan, 1995).
13. For a classic treatment of this issue see Benveniste, G., *The Politics of Expertise* (Berkely, CA: Glendessary, 1972).
14. Martin, B., 'Suppressing Research Data', *'Acountability in Research*, 6 (1999), 333–72.
15. Collinridge, D., *The Social Control of Technology* (New York: St. Martin's, 1980).
16. Smith, R., 'Reducing Violence'.
17. Alexander, 'The Future of Non-Lethal Weapons', 197–9
18. See Omega Foundation, *Crowd Control Technologies*, Report to the Scientific and Technological Options Assessment of the European Parliament, PEE 168.394 (Luxembourg: European Parliament, 2000).
19. See Davis, K., *Discretionary Justice* (Baton Rouge, LA: Louisiana State University Press, 1969).
20. See Rappert, B., 'The Distribution and Resolution of the Ambiguities of Technology, or Why Bobby Can't Spray', *Social Studies of Science*, 31, 4 (2001), 557–92.
21. Wilson, R., 'Representing Human Rights Violations', *Human Rights, Culture, and Context* (London: Pluto, 1997).
22. Ibid., 140.
23. Bhatt, 'The Extent of Problems and the Families' Perspective'.
24. Lindblom, C. and Woodhouse, E., *The Policy-Making Process* (Englewood Cliffs, NJ: Prentice-Hall, 1993).
25. For a number of detailed practical comments in this regard see Amnesty International, *Stopping the Torture Trade* (London: Amnesty International, International Secretariat, 1997).
26. Reitman, V., 'Seoul Adopts Chemical-Free Policy on Protests', *LA Times*, 5 March 2000, A-4 and Omega Foundation, *Crowd Control Technologies*.
27. Acres, P. 'Conflict Management Portfolio'.
28. Martin, B., *Social Defence, Social Change* (London: Freedom Press, 1993).

Select Bibliography

Ackroyd, C., Margolis, K., Rosenhead, J. and Shallice, T., *The Technology of Political Control* (London: Pluto, 1980).

Alexander, J., *Future War* (New York: St Martin's Press, 1999).

Altmann, J., *Acoustic Weapons* (Ithaca, NY: Cornell University Peace Studies Program, 1999).

Amnesty International, *Arming the Torturers: The Spread of Electroshock Technology* (London: Amnesty International, International Secretariat, 1997).

Committee on the Administration of Justice, *Plastic Bullets: A Briefing Paper* (Belfast: Committee on the Administration of Justice, 1998).

Council on Foreign Relations Independent Task Force, *Nonlethal Technologies* (Washington, DC: Council on Foreign Relations, 1999).

Council for Science and Society, *Harmless Weapons* (London: Barry Rose Publishers, 1978).

Dando, M., *A New Form of Warfare* (London: Brassey's, 1996).

Evans, R., *Gassed* (London: House of Stratus, 2000).

German Initiative to Ban Landmines and Landmine Action, *Alternative Anti-Personnel Mines* (London: Landmine Action, 2001).

Grint, K. and Woolgar, S., *The Machine at Work* (Cambridge: Polity, 1997).

Institut Chemische Technologie, *Non-lethal Weapons: New Options Facing the Future*, Proceedings of the 1st European Symposium on Non-lethal Weapons, 25–26 September 2001, Ettlingen, Germany (Postfach: ICT, 2001).

International Committee of the Red Cross, *The SIrUS Project* (Geneva: International Committee of the Red Cross Publications, 1997).

Jasanoff, S., Markle, G., Perersen, J. and Pinch, T. (eds), *Handbook of Science and Technology Studies* (London: Sage, 1995).

Kappeler, V. (ed.), *The Police and Society*, 2nd edn (Prospect Heights, IL: Waveland Press, 1999).

Lewer, N. and Schofield, S., *Non-Lethal Weapons* (London: Zed, 1997).

Lindblom, C. and Woodhouse, E., *The Policy-Making Process* (Englewood Cliffs: Prentice-Hall, 1993).

Maguire, M., Morgan, R. and Reiner, R. (eds), *The Oxford Handbook of Criminology* (Oxford: Clarendon, 1997).

Marine Corps, *Joint Concept for Non-Lethal Weapons* (Quantico, VA: Marine Corps, 1998).

Morales, F., 'Non-Lethal Weapons: Welcome to the Free World', *Covert Action*, 70, April–June 2001, 6–15.

Northam, G., *Shooting in the Dark* (London: Faber & Faber, 1988).

Omega Foundation, *Crowd Control Technologies – Technical Annex*, Report to the Scientific and Technological Options Assessment of the European Parliament, PE 168.394 (Luxembourg: European Parliament, 2000).

Prokosch, E., *The Technology of Killing* (London: Zed, 1995).

Rappert, B., 'The Distribution and the Resolution of the Ambiguities of Technology, or Why Bobby Can't Spray', *Social Studies of Science*, 31, 4 (2001), 557–91.

Robinson, J.P., 'Disabling Chemical Weapons', Presentation to PUGWASH Study Group on Implementation of the CBW Conventions, 27–29 May (Den Haag: PUGWASH, 1994).

The Sunshine Project, *US Armed Forces Push for Offensive Biological Weapons Development*, 8 May (Hamburg: The Sunshine Project, 2002).

UK Steering Group for Patten Report Recommendations 69 and 70 Relating to Public Order Equipment, *A Research Programme into Alternative Policing Approaches Towards the Management of Conflict*, December (Belfast: Northern Ireland Office, 2001).

Waddington, P. A. J., *The Strong Arm of the Law* (Oxford: Clarendon, 1991).

Wright, S., *An Appraisal of Technologies of Political Control*, Report to the Scientific and Technological Options Assessment of the European Parliament, PE 166.499 (Luxembourg: STOA, 1998).

Wynne, B., 'Unruly Technology: Practical Rules, Impractical Discourses and Public Understanding', *Social Studies of Science*, 18 (1998), 147–67.

Index